THEY ALL SWORE AN OATH

STORIES FROM EVERYDAY HEROES IN OUR MIDST

by

Christopher J. Lynch

ISBN-13: 978-0-9907273-6-1

AUTHOR'S PREFACE

This body of work is the culmination of a series of articles I wrote about veterans who were local to my communities. The articles were written for local newspapers - in Gardena, Ca. initially, and then, after my wife and I moved to Gallatin, Tn. for our local paper there.

The articles cover every branch of the service with the exception of the Coast Guard, and all races and ethnicities are represented. The time span runs from WWII, through Korea, Viet Nam, as well as Iraq, and the Cold War. The stories told run the gamut from the funny, insightful, daring, and heroic.

Not all of the subjects fought in combat, but they shared some other commonalities. To a person - whether drafted or volunteered, none of them regret their time spent in the military, and all consider it to be one of the most important periods of their lives. Following this point, they also were unanimous in their support of our country returning to a system of draft for military or other public service.

No matter what their respective MOS (job) in the military was: clerk, supply, machinist mate, or ground-pounding grunt, all swore an oath to uphold our constitution, even if it meant dying in the process.

We salute their service.

OATH OF ENLISTMENT
(Title 10, US Code; Act of 5 May 1960 replacing the wording first adopted in 1789, with amendment effective 5 October 1962).

I, _____, do solemnly swear (or affirm) that I will support and defend the Constitution of the United States against all enemies, foreign and domestic; that I will bear true faith and allegiance to the same; and that I will obey the orders of the President of the United States and the orders of the officers appointed over me, according to regulations and the Uniform Code of Military Justice. So help me God.

CONTENTS

LAST OF A PROUD LINEAGE

ANDREW AARON JR. MAJOR, U.S. ARMY

As fast and athletic as Trooper Andrew Aaron was, he couldn't get out of the way in time. A North Korean artillery shell hit the bridge he was standing on, blowing him into the air. He landed a short distance away, sustaining injuries to his shoulder and knee; shrapnel from the shell peppered his body. It was at that moment that the young man from New Orleans, Louisiana realized he was a long way from home.

The journey had begun several years earlier, in 1948, when Andrew Aaron answered the call to serve.

"The military had a recruitment drive going on at my high school, so I joined up. I was only sixteen and had to lie about my age, but I think they were just looking for warm bodies, so they didn't care."

Age was not a great area of concern for the military, but skin color was, and Aaron found himself and other African-American recruits segregated from the whites when he attended boot-camp at Fort Knox,

Kentucky. It was source of puzzlement for the young soldier.

"We would mix with the white soldiers from time to time, and I wondered why they didn't put us all together; we all seemed to get along fine."

Segregation notwithstanding, the young Aaron excelled in boot-camp and looked at it as a great competition. A former welterweight boxer, Aaron thought of the experience as one more great physical challenge in his life.

After graduating from boot-camp, Aaron was assigned to the 10th Calvary, one of the last remaining units of the storied "Buffalo Soldiers."

Established in 1866, the origin of the nickname "Buffalo Soldiers" has confounded historians, although all agree that it was a complimentary term given to the soldiers by the Native American tribes they fought against. Aaron was proud to be assigned to the legendary unit.

"I knew what it meant, and what the men who had come before me had done to earn the term."

At 17 years of age, he was shipped to Japan, and then before long, over to Korea. By now, the 10th Calvary had essentially been disbanded, and so Aaron was assigned to the 24th. Regiment, one of the original four Buffalo Soldier regiments. In short order he and his fellow Buffalo Soldiers headed into battle, in what everyone thought was going to be a short-term conflict. It was anything but, and before long, Aaron found himself deep in the mix.

"Combat is like a nightmare, and you realize real quick that you have to work as a team if you want to survive."

The 24th Regiment did much more than just survive; they saw the first victory in the Korean War. After a short stint in Japan to recover from his wounds from the artillery attack, Aaron returned to combat operations in Korea. By now, the military had become desegregated and the 24th was disbanded. Aaron joined his fellow African-American soldiers by becoming a part of the 2nd division. He received a Combat Commission, to 2nd Lieutenant in 1951. He spent a total of 19 months in Korea before returning to the states where he continued to serve his country, both as a US Customs agent, and an active Army reservist.

Trooper Andrew Aaron retired from the Army as a Major, with over 30 years of military service, including service during the Viet Nam War. Along the way, he found and married Geraldine Aaron, the love of his life, and together they raised a family. He was once honored to be California "Veteran of the Year" and has also held the office of President of the Los Angeles Chapter of the 9th & 10th (Horse) Cav. Association. He has dedicated himself to the development of the Aaron Community Cultural Center, a Buffalo Soldier Museum and Cultural Center in the Los Angeles area, where he hopes that this effort will inspire others from all backgrounds to achieve their goals, even in the face of adversity. www.buffalosoldiersonlinemuseum.com.

Author's note: Mr. Aaron passed away on January 17, 2019

THE MAN IN GREEN

JAMES McCARTHY, SERGEANT, U.S. MARINE CORPS

Without realizing it, the man in green had been working on young James McCarthy for the better part of half a decade.

"It was a poster that I had in my bedroom since I was 13 years old," McCarthy recalls. "He was a Marine and a tough one at that, a Force-Recon guy, all covered in green camo face paint. I thought he looked cool."

Cool or not, the image would take the better part of 5 years to have any effect on the young man from Cerritos, Ca. In the meantime, he was admittedly, a bit of a mess.

"I was a bad student during high school, and ditched all the time to go surfing – plus I hung out with some real unsavory characters."

In one instance, McCarthy realized that he was so far behind in his schooling, that he seized an opportunity to do a ride-along with the police to earn extra credit and get him out of the hole.

"It was fun," he remembers fondly. "But also, crazy. The night I did the ride-along, we busted some of the very guys that I hung out with. That was weird."

Whether from the boost that doing the ride-along gave to his grades, or by the skin of his teeth, he graduated high school and signed up to attend Cypress Jr. College, finding that he still had no direction in his life. That was when it hit him.

"I was taking basic classes, with no idea of a major or what I wanted to do in my life. One day, I was sitting in psychology class and realized that I had to make a change."

"I stood up and left class, went to the admissions office to drop out of school, and headed over to the Marine Corp Recruiting Station, where I enlisted on the spot."

The man in green had finally gotten to him.

1 month later he was at MCRD in San Diego for boot camp. And it was here that a bit of the old James McCarthy resurrected its mischievous self.

"Like all platoons, we had one guy who was dragging us down and getting the whole unit in trouble. One night, a couple of other guys and myself, "appropriated" everyone's shaving cream cans. We covered the guy in shaving cream while he slept."

At one point during the night, the victim woke up and, realizing that he was covered in shaving cream, leaped out of bed and ran to the DI's door, pounding on it. The DI opened the door, uttered one laugh, then got everyone out of bed to find out who was the culprit.

"We couldn't help but laugh, and so we fessed up to it," McCarthy said. "From that point on we were known as the "Shaving Cream Bandits," and we received a thrashing every chance the DI had to give us one."

The Shaving Cream Bandits did graduate from boot camp and McCarthy went on to his MOS training of Field Radio Operator in Twentynine Palms, Ca. It was a great job, and he really enjoyed it. Before long he was transferred to the 3rd. Recon Unit, 7th Comm division, where he handled satellite and microwave communication. It was here that his military career would take a decidedly different course.

"I met a Sargent who had seen duty in the MSG, or Marine Security Guard program, and it sounded like something I would really enjoy, so I applied for it."

McCarthy was accepted into the program and was sent to Quantico, Va. for training. The program was rigorous, with a constant cycle of PT, academics, and interview panels. The wash rate was over 80%.

McCarthy graduated the program, and felt very proud of being able to buck the curve. Now it was time to put the training to use by guarding one of America's many foreign embassies – a location that hardly could have been worse.

"I was sent to guard the embassy in Conakry, Guinea in West Africa, which was considered the second worst duty station in the world for MSGs. As soon as I stepped off the plane, I was given a flak helmet and a vest, but no weapon. I was to learn later that we could only have weapons on US soil – the embassy – but not outside. I was scared and nervous, wondering what the heck I had just signed up for."

He shared a house with 4 other Marines about 3 miles away from the embassy, and it was a real hot spot. The fence surrounding the house was 8 feet high, and the windows had steel louvers on them to keep out the gunfire that erupted on an almost nightly basis.

McCARTHY AT THE AMERICAN EMBASSY IN CONAKRY, GUINEA

McCarthy survived the duty station and excelled in his assignment, being promoted twice while in Guinea. He also was rewarded for his sacrifice by being transferred to Geneva, Switzerland to continue his service. It was a far cry from Guinea.

He mustered out of the service in 1996, and the day after he arrived home, he applied for LAPD, where he was accepted into the academy and served for 14 years. In 2010, James McCarthy transferred to the Gardena Police Department, where he serves to this day.

QUICK STUDY

MATTHEW MITCHELL, PETTY OFFICER 1ST CLASS, U.S. NAVY

Matthew Mitchell might have just set the record for the fastest military recruitment job ever. The Port Arthur, Tx. native had just left his local recruiting office after enlisting and stopped at the post office on his way home.

"I ran into a friend of mine there and when he asked me what I was up to, I told him that I had just joined the Navy, and that maybe he should to."

Texas in 1958 promised little in the way of jobs or opportunity, especially for an African-American, and so Mitchell's friend felt the same motivation as they turned around and headed back to the same recruiting station. The two men swore in, and were soon on their way to California and to the Navel Training Center in San Diego.

Normally, military basic training is considered one of the toughest things for a young man to endure, but not for Mitchell.

"I played football in high school in Texas," Mitchell recalls. "And let me tell you, when you practice football in Texas in the summertime, bootcamp in California is a piece of cake!"

The ease at which Mitchell breezed through the program must have caught the eye of his drill instructors, because he was soon promoted to RCPO, or Recruitment Chief Petty Officer. Essentially, he was now in charge of his fellow recruits.

"It meant that instead of the DI's doing all the yelling, I was doing it for them – and there was plenty to do."

Inspections for personal hygiene could make or break a platoon, and Mitchell had to make sure all of the men in his platoon passed muster – or they would all pay dearly.

"I carried a razor with me at all times, because some of the guys didn't want to shave." He laughs, "But believe me; they got shaved one way or another."

He graduated from bootcamp 3 months later and a whole lot heavier, going from his entry weight of 151 pounds all the way up to 180. From here it was on to Great Lakes Machinist School in Chicago, Illinois where he graduated as a Class A Machinist, the only black in a class of forty whites. He was assigned to the USS Kidd and worked to overhaul the ship before heading out on a West-Pac tour of the Pacific.

MACHINIST-MATE MATTHEW MITCHELL

"It was really great seeing another part of the world, and I saw some beautiful scenery and sunsets while we were out there."

But he also saw things that weren't so idyllic.

"It was only 1959, and most people had not heard of Viet Nam, or knew about the growing conflict in this part of the world, but we watched explosions from the deck of our ship as shells rained down on the Matsu Islands."

Although Mitchell wasn't in harm's way from his vantagepoint on the ship for that incident, there were plenty of other times when the machinist mate felt he might just meet his maker.

"I've been through two typhoons, and they were scary. Imagine two hundred hurricanes altogether."

One time, his ship was entering a port and they had to pass through a narrow strait with tall cliffs on either side. An enormous wave hit the ship and literally picked it up and moved it 500-600 yards. Mitchell was manning his station down in the engine room and was holding on for dear life.

"I don't know how big that wave was, but all I know is that I had quite a sinking feeling in my stomach."

After being transferred to serve on the Henry W. Tucker, DDR-875, Matthew Mitchell was discharged from the Navy in May of 1962 and returned to California where he took a job at the shipyards and climbed in the ranks once again. He became a ship scheduler, and went on to earn a BA in industrial arts with a major in engineering drawing. He also earned an MA in education, and by the time he retired had achieved the rank of supervisor of production engineering.

He met his wife in 1969 and together they had two daughters and eight grandchildren.

His wife passed in 1994 and today, Mitchell spends a good portion of his time working with his American Legion Post 109, which, he is told, is the only faith-based Post in the United States.

A TALE OF TWO MARINES

RODNEY GONSALVES, 1st SERGEANT, U.S. MARINE CORPS RESERVE

Not many Marine recruits have a sense of Deja vu when they step off of the bus on the first day of boot camp; but then again, Rodney Gonsalves was not your average Marine.

"My father died when I was very young, and my mom was trying to raise my two younger brothers and me by herself - which was very tough. Then she found some assistance from a very unique program."

The program was called the Young Marines, and it took boys all the way down to seven years of age and molded them into responsible young men in much the same way as the military branch did, through discipline, pride, and by fostering a sense of responsibility. Gonsalves, twelve years old at the time, couldn't wait to join.

"Our drill instructor was Sergeant Souza, who was instrumental in starting the Young Marine Program and had been in the Marine Corps during the Korean Conflict. He was very strict, but he was very fair."

The Young Marine boot camp lasted for approximately one year, and Rodney and his two younger brothers; one aged ten, the other eight, all graduated successfully. Rodney had been recognized as the outstanding graduate of the cycle, and for his exemplarity performance, he was given his own platoon to command.

"At the tender age of thirteen," he recalls proudly. "I became a junior drill instructor."

RODNEY, UPPER LEFT, WITH HIS MOM AND HIS TWO BROTHERS AT THEIR JR. MARINE GRADUATION

Besides training and drilling at their meetings, the Young Marines also were able to go on some very exciting field trips throughout the years.

"We visited West Point, Camp Lejeune, and then at one point, we got to go to Parris Island, South Carolina. I loved it, and it was then that I realized that I wanted to come back here again. I wanted to experience the real Marine boot-camp, and to become a Marine."

He left the Young Marines at the age of sixteen to concentrate on his studies, and then in 1979 joined the Marine reserves and returned to Parris Island.

Having been through the Young Marine program for several years had prepared him well, and he pretty much breezed through boot camp. As easy as it was on him compared to the other recruits though, the DI's still enjoy finding a chink in the armor and exploiting it.

"We were standing on line and it was a hot day and so we had our shirts off. The DI was strutting up and down the ranks and talking and when he got to me, a fly landed on my bare chest. The DI and I both knew that I wanted to swat it, and so he egged me on."

"You want to swat that fly, don't you Recruit?" the DI asked.

"I knew that I couldn't, and so I said, 'no sir.'"

Then the fly started to bite Gonsalves - hard, and blood began to trickle down his chest. The DI continued unabated.

"Now you REALLY want to swat that fly, don't you Recruit?" he asked again.

"No sir."

The fly finally flew off and the DI moved on. Gonsalves survived the incident, and avoided what would have been a quick date with "incentive physical training" for giving the wrong answer.

Biting insects notwithstanding, boot camp also provided some more enjoyable experiences and one that Gonsalves could have scarcely imagined.

GONSALVES, BOOTCAMP

“We were out training at a remote part of the base when all of a sudden, I saw a bus filled with Young Marines pull up – along with my old drill instructor and mentor Sgt. Souza! It took some convincing to my Senior DI, but I was finally able to leave my platoon momentarily to go say thanks to the man who had been such an integral part of my life. What a reunion it was, even if for only a brief moment"

After boot camp, Gonsalves was trained as an air traffic controller and he stayed in the reserves until 2002. In the meantime, he joined the Gardena Police Department where he is an Administrative Sergeant. He is also in charge of the department's Honor Guard, as well as teaching drill to the Police Explorers.

A few years later, Gonsalves' Young Marine Drill Instructor, Sgt. Souza passed away. Rodney attended the funeral proudly in his dress blues and was taken aback when Sgt. Souza's son asked him to stand by the casket until the cemetery workers could lower it into the ground.

"It would be an honor and I'll take good care of him," Gonsalves quickly responded.

He stood at parade rest guarding the casket until the workers arrived a short time later. And as they lowered it into the ground, First Sergeant Rodney Gonsalves gave a final salute to the person who had done so much to mold him into the man he is today.

COMIC RELIEF, OR…HOW TO BRING AN AXE HANDLE TO A GUN-FIGHT

GUY CERDA, CORPORAL, U.S. ARMY

Guy Cerda was in a deep slumber when his life changed abruptly. The eighteen-year-old Los Angeles native was working the night shift at a local supermarket, and was sleeping soundly when his mother shook him awake on that fateful morning in January 1970.

"I think you better have a look at this," she said.

Cerda rubbed the sleep from his eyes and stared at the letter his mother was holding in her hand. It was from the Selective Service System; He was being drafted.

"This was mid-January and I was to report for induction on February third. It instructed me that I essentially had two weeks to get my affairs in order."

Cerda admitted that he was scared, but also that he had no hesitation about showing up.

"This was during the time that a lot of guys were dodging the draft and heading to Canada to avoid being inducted. But Uncle Sam was calling me, and I thought; Okay, I'm going to do my commitment."

A few short weeks later, Cerda was one of a number of nameless, faceless recruits doing basic training at Fort Ord in Northern California. It was a frightening, life-altering experience, but one that the Jefferson High graduate soon reveled in.

"I thought it was a fantastic experience, and one that every young person should have. It really reemphasized the good qualities we as individuals should have: discipline, respect for other people, for property, and for authority."

Besides contributing to his mental maturity, boot camp also developed Cerda in ways he never could have imagined.

"When I entered boot camp, I was a scrawny one hundred forty-five-pound kid. And when I graduated, I was a one hundred sixty-five-pound man; between all of the exercise and square meals, I packed on twenty pounds of muscle in just eight weeks."

It was also in boot camp though that Cerda experienced one of the most comical moments in the service.

"We were only three weeks into basic when myself and a couple of other recruits were ordered to pull guard duty and protect the ammo dump. The only problem was, we hadn't been qualified yet with firearms and so they couldn't give us guns. So instead, they gave us axe handles and told us to guard the dump from any intruders. I don't know what we were supposed to do with axe handles if someone really wanted to break in, but in the military, you do what you are told to do with what you have."

After basic training it was determined that Cerda would stay stateside to help serve his country. His experience working in a grocery store had made him a natural for a job as a supply specialist. He remained at Fort Ord and was assigned to the 101 St. Transport Unit. He and his fellow soldiers were responsible for maintaining the stocks that a fighting force would need to be mobilized anywhere in the world within forty-eight hours.

"We didn't have any specific training classes, and instead learned our jobs through OJT."

Cerda must have learned well, because during his stint at Fort Ord, orders came up twice that he was to be deployed to Viet Nam.

"My superiors always pushed back though, and said that I was too valuable for them to let me go; they always got my orders waived."

Despite his busy schedule, Cerda still found time to carry on with the part of his life he left behind when his draft notice came through.

"I knew that there was always a chance that I would get sent to Viet Nam, and I really wanted to have something to come home to and to fight for, and so I got married while I was serving. And before long my wife was pregnant with our first child."

Guy Cerda was mustered out of the Army after just twenty-one months, three months shy of his two-year commitment. President Richard Nixon, the same Commander in Chief who had ordered the randomizing of the Selective Service through lottery, was now planning to eliminate it altogether and draw down the troop levels.

He left active duty, but continued to fulfill his commitment of four years as a reservist. In the meantime, he attended college and built a comfortable life for he and his family.

Although he never saw combat, Guy Cerda is proud of his military service, and would have gone into battle if ordered.

"The one thing I know," he says. "Is that this freedom is not free, and that someone has to pay for it."

PART OF SOMETHING BIGGER THAN HIMSELF

LOUIE SCHWARTZ, GUNNERY SERGEANT, U.S. MARINE CORPS RESERVE

Louie Schwartz stood staring at the side of the refrigerator, unable to believe his eyes.

"It was my report card from our local community college, and it was straight A's. It was the first time in my life that my father had ever been proud enough to display it - I was twenty-three years old."

Louie Schwartz credits his stint in the Marine Corp for changing his outlook on life, and for making him realize his full potential.

"Up until this point in my life, I hadn't been the best student, but I knew that I had to have good enough grades to achieve my dream, which was to serve in the USMC."

And so the California native did just that, maintaining his GPA while serving in the Redondo Union High School ROTC. He enlisted in the United States Marine Corp. immediately after graduating in 1994` and was soon on his way to bootcamp at nearby MCRD in San Diego.

He graduated from bootcamp in January 1995 and headed up the road to Camp Pendleton's SOI

(School of Infantry) to begin his specific MOS training.

"I was initially slated for basic infantry, but after graduating as the honor graduate with my class, I was offered a chance to become an anti-tank and demolition "assault-man."

He and another Marine were the first two graduates from his assault-man class to be assigned to the battalion in that capacity, but Louie Schwartz wanted to set his sights even higher.

"I applied for recon training with the 1st Recon Company and was accepted and put through 'The Pipeline' as it was called."

The pipeline consisted of some of the hardest training the young Marine had been through: amphibious assaults, land navigation and rappelling. It was, in his words, "pain, misery and agony," every day – but he reveled in it, and considered his fellow Recon Marines to be the best he had ever been associated with.

He left active duty with the Marine Corp in 1999, but stayed on as an active reservist. In the meantime, he enrolled at El Camino College in Torrance, Ca. where he made the dean's list before moving on to UCLA.

He wouldn't be a Bruin for very long before being called up to serve in Operation Iraqi Freedom in early 2003.

RECON TEAM 1ST FORCE NEAR THE IRAQ BORDER WITH IRAN. SCHWARTZ IS 2ND. FROM RIGHT.

"Being with recon, it was our job to try to clear the path for the troops that were going to be coming in for the invasion. We would skirt around Baghdad in the middle of the night and locate potential threats to our troops."

Schwartz never considered himself special for being in a Reconnaissance Unit.

"We don't win the wars," he says humbly. "The infantry guys do. It's just our job to try to let them win it with the fewest casualties."

Schwartz returned after a seven-month deployment and continued his studies at UCLA, where he graduated in 2004. In 2006, he joined the Gardena Police Department while still maintaining his active duty reserve status. Becoming a member of the Gardena Police Department was, in his words "One of the greatest moments of my life."

AS A GARDENA POLICE OFFICER

He was called up again in 2007 for another seven-month deployment in Iraq, this time as a joint tactical air controller, whose job it was to find threats to our troops from the air. It was here that Schwartz had his biggest find and the one that he is most proud of.

"I noticed an anomaly with an image of a road our troops were going to be travelling on. We had a special unit check it out and they found that it was an IED with a delayed trigger mechanism that would have taken out two squads of men. I was glad I was able to spot it and to save them."

JTAC TRAINING EXERCISE

Gunnery Sergeant Louie Schwartz retired from the reserves in December of 2015. He currently works in the intelligence community.

HAPPY BIRTHDAY, RECRUIT

RYAN WILKES, SERGEANT 1ST CLASS, U.S. ARMY

Recruit Ryan Wilkes could hardly believe his good fortune. For weeks the drill instructors here at Fort Benning, Ga. had been all over him: in his face, screaming at him, pushing him beyond his physical limits. Now, here they were, merrily celebrating his birthday with him. The entire platoon sang *Happy Birthday* to him, and the DIs even got him a birthday cake to eat.

"They told me to have seconds, even thirds of the cake if I wanted to," Wilkes recalls.

Then the reality hit.

"As soon as I was done eating, they ordered me to put on my MOPP-4 gear, or *Mission Oriented Protective Posture*. It was essentially a fully encapsulated biological warfare suit – then they had me start running laps, until I vomited."

Happy 18th birthday, and welcome to the United States Army.

It had been quite a journey for the lifelong Gardena, Ca. resident, and one that wouldn't be over soon. And there was no time to reflect on what the DIs had done to him in bootcamp. With his grasp of mathematical skills such as trigonometry, he became a scout/mortarman, and found himself moving up fast in the ranks.

"I made Sergeant by the time I was nineteen, only eighteen months out of basic, which was almost unheard of, "Wilkes states.

After a short stint in Vilseck, Germany, he was sent to Fort Polk, La. and continued there until his three-year contract was nearly up.

"They wanted me to re-enlist, but I told them I would only do it if I could go airborne."

As a child growing up, Wilkes would watch all of the WWII movies with his Grandfather, a veteran of the Army-Air-corp. He thought that the parachutist of the era had one of the most exciting and important roles in the military, and he wanted to be one.

After a few miscues, he finally got his wish and was assigned to the I-509 Parachute Infantry Regiment. But there were no openings for jump-school, and he would have to wait to get his "wings." Until then, he was low man on the totem pole all over again.

"If you're in an airborne unit and you don't have your wings, you are known as a "leg" and you are tacitly outranked by everyone below you who does have their wings. The hazing in there was brutal."

Finally though, a spot opened up and Wilkes found himself back at Fort Benning for jump school. He still remembers the day he graduated, standing in a long line of airborne wanna-be's waiting to make their first jump.

"I remembered standing there, thinking, 'I'm going to jump out of an airplane.'"

Before he had a chance to ponder his situation further, the jump light turned green and he headed out the back of the plane and into the wild blue yonder.

There were many jumps after that initial one to earn his wings, and they weren't the only wings he earned. The next pair he pinned on were from one of the Unites States' longtime nemesis.

"I was able to earn Russian wings as well," Wilkes says. "We trained with the Spetsnaz, or Russian Special Forces."

Under the direction of a Russian Jumpmaster barking commands in Russian, Wilkes and a select group of soldiers parachuted out of an Antonov An-2 Colt, a Russian bi-plane that traces its roots to the end of the Second World War.

ANTONOV AN-2 COLT

The game was about to turn deadlier though, as Wilkes deployed in February of 2004 to Iraq. In a little more than half a year into his deployment, his Humvee was hit on August 15th, 2004. Wilkes was critically injured in the attack and lost half of his large intestine.

He endured several operations to patch things up, but eventually was given a medical retirement as a Sergeant 1st Class in 2005. Today, Ryan Wilkes continues to serve our community, as an officer with the Gardena Police Department.

BIG GUNS, BAD EYES

SCOTT TAKAHASHI, SERGEANT, U.S. ARMY

As the artillery shells from his battery flew overhead that first night of the fire mission, the only thing that Scott Takahashi could think of was, "I'm never going to be able to sleep for an entire year."

Takahashi did learn to fall asleep though, through the deafening noise and the dirt that fell from the ceiling onto him in his hooch with each 155 round that sailed overhead. He learned to sleep and to adapt to his new surroundings; Battery A of the 2nd Battalion 35th Artillery, deep in-country in Viet Nam, even though he wasn't even supposed to be there.

This wasn't a *Private Benjamin* moment for Takahashi, as in the scene where Goldie Hawn thinks she has somehow been sent to, "the wrong Army." He knew he was in the right army, he just didn't think that someone with his physical deficiency could be sent to a war-zone, especially to aim giant guns that could inflict incredible amounts of damage.

"My dad had told me when I got drafted: 'don't worry; with your eyesight, they'll never send you to Viet Nam."

It was November 1968; the height of hostilities in what had been formally known as French Indo-China. The Gardena, Ca. resident was attending LA Trade Tech and Harbor College when his draft notice arrived.

"My dad had served in WW II and fought for our country, and so I knew instinctively that living in America is a privilege, and so when you are asked to go – you go."

And so with the reassurances of his father's comment about his eyesight ringing in his head, he arrived on his assigned day at the induction center for his physical and processing.

"When they removed my glasses and asked me to tell them what was on the eye chart, all I could say was that I knew that there was an "E" on the first line – and that was from memory. But they took me anyway."

He "celebrated" his 20th birthday in the reception area of Fort Ord, Ca. before moving to the barracks to begin his basic training.

"They put us all into the back of a big truck to take us there and when we stopped and they yelled for us to get out, I jumped out of the truck with my duffel bag and promptly knocked over a drill instructor!"

"Great," I thought. "We haven't even started basic training and already I'm screwing up. It cost me twenty-five pushups, the first of many."

For the next several weeks of basic, and following that with his training as a supply clerk, Takahashi not only made up for his first Faux Pas, but proved himself to be a very capable soldier. Impressive enough that after all of his training as a clerk, he was going to serve a higher calling.

"I was supposed to be going to Fort Sam Houston until one day an executive officer walked up to me with a big grin on his face to let me know I was going to Viet Nam. So much for my dad's theory of military fitness."

At Bien Hoa, he was assigned to an artillery unit, even though he had no training with the heavy guns, or even on the hazards of the country itself. He worked with the unit in Bien Hoa for 3 weeks before being sent out into the field for what would amount to many missions over the next 13 months.

"We were in the field about 75% of the time, and I soon learned that artillery and humping 155 shells was tough work."

But things were about to get a lot tougher.

"In April 1970, we were hit hard by the 33NVA Regiment."

It was vicious battle, and Takahashi saw plenty of combat manning the berm with a machine gun. For his bravery in that fight, he earned the Bronze Star for valor.

SCOTT TAKAHASHI RECEIVING THE BRONZE STAR FOR VALOR

Decades after that battle and the war had ended, and Takahashi had returned home, he is still active with his battery and has organized reunions for the surviving members. He's also the Vice President of the 100th Battalion Club and also works with The Go For Broke Education Center, a group dedicated to educating the public about the responsibilities, challenges, and rights of American citizenship, by using the life stories of the Japanese American soldiers of World War II.

Reflecting on he and his father's service to our country, Takahashi has no regrets. He's proud of he *and* his father's service and says that, "Combat forges bonds that are unbreakable. Sometimes we who served may disagree about exactly how things happened, but it does nothing to diminish that bond."

WAR AND PEACE (CORP)

TERRY KENNEDY, SPEC-4, U.S. ARMY,

The easiest thing for Terry Kennedy to do would have been to head north to Canada. After all, he was born there and had plenty of relatives still living up there.

"My dad really encouraged me to do it," Kennedy recalls. "My brother had been drafted ahead of me, and had done a tour in Viet Nam. I guess my dad really didn't want to go through that again."

But Kennedy knew that he couldn't dodge the draft; he had become a US Citizen many years earlier, and had sworn his allegiance to the country that he now called his own. The year was 1967 when he received his draft notice. It couldn't have happened at a more tumultuous time for the US, and for Kennedy, under more incongruent circumstances.

"I was in Puerto Rico at the time, teaching Spanish to, of all people, Peace Corp trainees."

Kennedy had volunteered for the Peace Corp three years earlier in 1964 when he graduated from college. At the time, there was a belief that if you were in the Peace Corp, you weren't subject to the draft.

Not true.

The bogus Peace Corp exemption notwithstanding, Kennedy thought he might never have to serve for another reason: he was getting too old.

"I was 26 when my notice came, and the cutoff was 27, so I almost made it," he laughs.

And so it was off to Fort Ord in Northern California, where Kennedy not only went through basic training, but also found that he was the oldest recruit in his platoon. Being the eldest didn't help allay any of his fears though, and it was admittedly, a very scary time for Kennedy.

"We all knew what the odds were about going to Viet Nam and possibly getting killed, so it was always in the back of your mind."

He was trained as a clerk, and because of his college degree, was soon tasked with teaching others in his unit, an assignment he thought might keep him from going overseas.

Again, not true.

"My entire unit was shipped to Korea. I was the only one who was sent to Viet Nam. So much for luck."

He arrived shortly thereafter and was sent to First Air Calvary Division at the DMZ, where his job was to type in the coordinates for B-52 bombing runs. After three months on the DMZ, he was sent back to a base north of Saigon, where the fighting was equally as fierce.

"Some of the bombing coordinates I had to send out were very close to our base, and you could feel the earth shake when the bombs hit."

They also took incoming fire regularly, and even as a clerk he learned, among other things, how to identify the distinct "whistling" sound of an incoming missile. The Viet Cong were terrible shots, but every once in a while, they got lucky.

"A hooch on the base took a direct hit one night, and everybody inside got killed," Kennedy said. "After that, we had sandbags all around our cots when we slept."

Despite the constant danger, there were some lighter, more humorous moments as well.

Kennedy smiles. "A lot of the guys in my unit, they were rougher than I was, and they noticed that I never used profanity - ever. Well, this was something that they could not tolerate, and they began working on me. Finally, one night I had had enough and said, 'Okay! What do you want me to say?'

"They told me that I had to use this one word, a very foul word — in every way I could possibly use it, for the next five minutes. And, so I did.

Then, they were satisfied."

Kennedy returned to the states in 1969 and settled in California, where he still lives today with his wife.

UNSINKABLE

JAMES (PAPPY) KEMP, SEAMAN, U.S. NAVY

Seaman James Kemp was on watch, stationed on a catwalk on the starboard side, when the first torpedo from the German U-Boat, U-549, slammed into the port side of the ship.

The force of the blast threw him over onto the catwalk. And when he looked back up, he saw the fuselage of an aircraft hanging precariously above him, ready to fall. He knew then that he was a long way from home.

Born in Nashville, Tn. in 1926, Kemp lost his father at an early age and was raised by his grandparents in Smith County where he attended high school. It was a part of his education that would be cut short as WWII raged around the world, calling to young men such as Kemp to serve their country. He enlisted in the Navy over the Christmas break, never going back.

Kemp recalls, "I never saw any good reason to return to school, so I didn't. I just waited to ship out"

On his eighteenth birthday in February 1944, that's exactly what happened as he was sent to basic training in Maryland.

The term 'basic training' is operative here as young men were given only the most rudimentary of training before heading into the theater.

Kemp was assigned to the CVE-21, the USS Block Island, and before he knew it, was steaming out into the Atlantic Ocean.

The Block Island was one of a unique set of warships. Converted from tanker hulls, they were part of an escort class of carriers designed to protect merchant ships as they crossed the Atlantic to bring war materials to Europe.

At only 500 ft. in length, the ships – nicknamed 'Baby-Flattops' - were smaller than regular carriers, but still carried a compliment of about two-dozen aircraft. The planes served dual purposes; on the trip over they flew reconnaissance missions to watch for the deadly German U-boats that prowled the seas, and at the termination of the voyage, would be delivered to Europe to resupply pilots over there.

USS BLOCK ISLAND

With the Atlantic teaming with deadly German U-boats, Kemp saw action only a few months out to sea. Along with the compliment of destroyers that made up the escort task force, they had identified and then disabled the German U-801.

The submarine surfaced, but was unable to resubmerge and slip away. Kemp and his crew took the German crew onboard as POWs.

"We put them into the bow of the ship to hold them as prisoners," Kemp remembers. "And they were actually pretty well behaved."

They sailed with the compliment of POWs to Casablanca, where they dropped them off. It was a place that Kemp would soon revisit for a different reason.

It was 8:13 pm on the fateful night of May 29th, 1944 when the first torpedo slammed into the port side of the Block Island. Kemp recovered his footing on the catwalk and moved away from the threat of the dangerously perched plane above him, just as a second torpedo hit.

It was then that the captain gave the order to "man your battle-stations."

To Kemp, this meant moving over to the port side of the ship and manning a 20 MM gun. He had only been on the gun for a few minutes when the third - and final – torpedo struck the port side of the ship. The force of the explosion threw Kemp into the big gun face-first. He lacerated his forehead, and received two black eyes in the process.

With the Block Island now hopelessly damaged, the order came to abandon ship.

USS BLOCK ISLAND SLIPPING BENEATH THE WAVES

Kemp remembers standing at the rail wearing his "Mae-West" life jacket. He could have jumped into the water then, but waited.

"I wanted to make sure my buddies got into the water first. I didn't want to see them get left behind."

Finally, when it was clear that his friends were safely in the water and swimming toward the rescue ships, Kemp took the plunge himself.

He swam toward the rescue ship, over 500 yards away, battling not only the current, but the oil and gasoline on the surface that saturated his clothes.

"I was so exhausted when I got to the ladder on the other ship, that I could hardly climb it," he recalls painfully.

The Block Island slipped beneath the surface in only minutes, becoming the only US carrier to be sunk in the Atlantic during WWII. Thankfully, of the compliment of 957 men, only 6 died.

The surviving crew members were taken to recover in Casablanca, before heading home for thirty days of survivor leave.

After the incident, a new Block Island (CVE-106) was commissioned and along with the majority of his crewmates, Kemp, served on it in the Pacific Theater, until the war's conclusion in 1945.

He returned to Gallatin, Tn. where he soon married and began a long and colorful career, first as a highway patrol man, then as a Federal Marshall. He was even recognized by the White House in 1962 for being part of a contingent of Marshalls that protected

James Meredith as he became the first African-American to attend Ole Miss College in Oxford, Ms.

Author's note: James "Pappy" Kemp passed away on Oct. 14th. 2022.

CARRY ON

DARRYLL A. HOLGUIN, SPEC-4, U.S. ARMY

On the first day of Bootcamp in Polk, Louisiana, Darryll Holguin realized that he had stepped into a different world.

"They were mostly California boys in the group I had been sent there with, and being that it was the sixties, most of us had long hair," he recalls. "And so, before they even began to process us, the DI's decided to have little fun with us at our expense."

Holguin and his other longer haired recruits were marched around the parade deck, much to the entertainment and cat-calls of the seasoned soldiers there. Welcome to the United States Army.

The year was 1968 and the Santa Ana, Ca. native had received his draft notice and was inducted on April 29. Even though there were no draft lottery numbers yet, (they would begin on Dec. 1st. 1969) it wasn't a surprise to him.

"The war was really heating up," he said. "And you saw all these other guys you knew getting drafted, so you were pretty much expecting it."

Despite the less than warm welcome, Holguin loved Bootcamp and the physical challenges it presented.

"I gained 15 pounds in Bootcamp, and it was all muscle." He says proudly, then adds, laughing, "In fact, my own brother didn't even recognize me when he came to pick me up, and kept driving around and around the block looking for me."

What's uniquely odd about his brother not recognizing him was that fact that his brother had served in the military himself. In fact, *all* of Holguin's 6 other brothers had served in either the Air Force or the Army.

After graduation he was sent to Fort Rucker, Alabama to begin work in his assigned MOS, 64B, Heavy Duty Truck Driver. Although he admits that he mostly drove a jeep around escorting warrant officers, there were some interesting times too.

"We would take the officers out to the field to watch the helicopter pilots learning to fly, and when we had a chance to fly on one of the choppers with them, my buddy and I jumped at the chance. Pretty soon it became a routine for us and we were flying all the time. The instructors thought we were nuts for wanting to get into a helicopter with a trainee pilot!"

IN COUNTRY, IN THE 5 TON

It was at Fort Rucker that Holguin also learned he was going to be sent to Viet Nam, *and* where he was presented with an opportunity to go AWOL and desert.

"I had gotten to know some guys in the local town and when they heard I was going to be sent to Viet Nam, they told me they could help me escape to Canada."

Holguin refused to desert the country that he loved, and was soon '*in-country*' where he was assigned to a combat unit to drive a water truck and replenish units in the field. It was here that he endured one of his first close calls with the enemy.

"I had just walked from the truck and was about 20 yards away when a mortar round hit right near it," he remembers grimly. "The explosion knocked me out and knocked me to the ground. It was scary."

And if a water truck was a target for the enemy, his next assignment, ammunition handler, made him an even more attractive one.

"I delivered all sorts of munitions to the guys in the field, Claymores, 42mm mortars, M-79 grenades, .50 cal — *everything*."

While performing these duties, Holguin then had another close call.

"I was unloading mortars off the truck and wasn't wearing my helmet because it made it so much more difficult. A sniper took a shot at me then and he was so close that I felt the bullet go through my hair. A couple of inches lower, and I wouldn't be here."

RECEIVING THE BRONZE STAR

These weren't the only frightening experiences he endured during combat and found that he would pay for it much later on in life. After being sent back home, discharged, and settling into civilian life with a family and children in California, he found that he was easily agitated and as he explains it, became a workaholic as a way of coping. At one point, he realized he needed help.

Treatment at the Long Beach, Ca. VA was long and painful, but he eventually got through it and acknowledged his PTSD. But after realizing how many others were suffering right along with him, he decided to carry on the fight. He and some other vets decided to start an advocacy group to help their fellow brothers and sisters in arms. The organization is called the, "Veteran's Voice, LB Council" and, along with the endorsement of the hospital board, have helped numerous veterans navigate the often frustrating and byzantine works of the VA. Holguin volunteers with the group at the VA several times a week and has made it one of his life's ambitions, realizing he is helping himself as much as others. You can read more about the organization or contact them at their website: veteransvoicelb.org

GROUP EFFORT

BILL HORN, SPEC-5, U.S. ARMY

BILL HORN, (LEFT) BEING HONORED BY THE GARDENA CITY COUNCIL

Bill Horn and his Ukulele group, *The Rock-a-Lay-Lees*, stand before the crowd at the Long Beach Veterans Hospital and take in the warm round of applause. Many of the vets are infirmed or crippled, and some are even blind. The *Rock-a-Lay-Lees* have performed here at least a half dozen times before, but for Horn, a former Army Spec-5, bringing a bit of joy to his fellow veterans never gets old.

"I always feel a special kinship to them," he admits. "Even though their circumstances are worse than mine."

The bond he feels was forged a long time ago when in 1966, Horn, then just barely 18, stood with hundreds of others in the downtown Los Angeles Induction Center and recited the Oath of Enlistment for Military Service. It was a special moment for the young man from El Centro, Ca.

"There was a lot of pride for me taking the oath," Horn recalls fondly. "I knew that I was going to be serving our country, and that was very special to me."

After the oath, it was a several hour bus ride up to Fort Ord in Northern California for basic training. It was, admittedly, tougher than Horn had thought it would be.

"We had an exercise where we had to crawl on our bellies through mud and under barbed wire, while all the while, machine gun rounds were being fired above our heads. To make matter's worse, I had a full pack and had a weapon that I had to try to keep above my head so that it didn't get dirty – and, I was sinking in the darn mud!"

The young recruit made it through that exercise as well as the rest, and it was also in boot camp where he learned about the comradeship of his fellow brothers in arms.

"We had a hump of about 15 miles with full gear and weapons, and guys were falling out left and right. I was determined to make it but about a mile from the end, I slipped on a trip root and sprained my ankle. Two other recruits helped me the rest of the way and together all three of us made it to the finish."

From Fort Ord, he headed to Fort Bliss, Tx. to train in his MOS (25-G), as an electronics technician supporting a Fire Unit Integration Facility. The 'facility' essentially amounted to a large trailer with data processing equipment linked to miles of cable that went to Hawk Missile and artillery batteries that they were going to trained to operate as well as repair.

Bill and the others in his class, all 9 of them, passed the 8 week AIT course with flying colors. They were so good that upon graduation, they were assigned to be instructors to train others in the same skills. It was an experience of international proportions.

"Germany had the Hawk Missile, as well as Italy, France, Belgium – even Japan, and we had to train all of them."

The classes with the students from far-flung countries ran 24 hours a day which put Bill and his fellow instructors in the position of having to often present training at 2 o'clock in the morning. Still, his sense of duty to mission and teamwork got him through it.

Even as classroom bound instructors though, they still were expected to requalify with their weapons and this led to one of the most comical – albeit scariest, moments in Horn's military service.

They were on a firing range with dozens of others and after requalifying from the prone and standing positions, were tested in the 'foxhole' position, a situation where Horn and two others had to run and jump into a foxhole, ready their weapons, and fire. They performed well enough to requalify, but it was when they were leaving the hole that things went very bad.

"We were told by the range master to cease fire, lock our weapons, put them on 'safe' and shoulder them before climbing out. Myself and another guy were doing what we were told and were shouldering our weapons when this other guy, a smart assed guy from New York who we called 'Jack-Wagon', jumps out of the hole, his weapon not locked, not on safe, and with around in the chamber and then – BAMM!"

The round discharged and went down the firing line, striking a piece of metal on the end.

"The scorer looked at the range-master in the tower, the range-master looked at him and yelled, 'Get that man!'" Horn adds ruefully, "That was the last we ever saw of Jack-Wagon."

THE UKULELE GROUP, ROCK-A-LAY-LEE WITH BLIND VETERANS AT THE LONG BEACH VA HOSPITAL

Bill Horn mustered out of the US Army in August of 1969 after having been awarded the National defense Service Medal, the Expert Rifle

Qualification Badge, and the Good Conduct Medal. He built on his military electronics training by pursuing a career in the burgeoning aerospace industry in Southern California, working for NCR, Hughes, and then finally Raytheon. Along the way, he married and between he and his wife have 4 children, 16 grandchildren, and 8 great grandchildren.

His Ukulele group still performs at the Long Beach VA, and they hope to soon begin entertaining vets in transit at the LAX USO.

BUCKING THE ODDS

RUSSELL REEVES, PRIVATE 1ST. CLASS, U.S. ARMY

The ratio wasn't good: twenty unshackled German POWs, one newly minted GI, and one M-3 'Grease-Gun' with a thirty-round capacity – barely more than one round per POW. To say that Army Private Russell Reeves felt somewhat intimidated, would be an understatement.

Born and raised in Paris, Tn, Reeves and his family had barely heard the news that his brother had been killed attempting to retake the Philippine Islands from the Japanese in December 1944, when his draft notice arrived. He listed Navy as his first choice, but the US War Department had other ideas.

He was inducted into the army and sent to Camp Gordon (currently Fort Gordon) in Augusta, Ga. for basic training. It was here that he was informed that because of his age and fitness for duty, he would be going to the infantry. Besides the specter of combat hanging over his head, Reeves describes the experience as a miserable one; hot, humid, and bugs, bugs…BUGS!.

"I bet I ate a ton of gnats while I was there," he laughs.

RUSSELL REEVES, HOME ON LEAVE

After graduating from bootcamp and having two weeks of leave, he was sent over to Le Havre, France via the transport ship, the USS West Point, on what would be its final European voyage. By now, August of 1945, Hitler was dead and Germany had surrendered, but Europe was in shambles and needed to be rebuilt.

Along with several other GIs, Reeves was transported from France to Belgium by train in a freight car that was affectionately known as a, 'forty-plus-eight' for its capacity.

"The 'forty-plus-eight' moniker comes from the fact that it could transport forty men, and eight mules," Reeves explains. "Thankfully, we didn't have any mules to contend with. But when one of the guys opened the side door to get some air in, it fell right off the car and into the weeds."

In Belgium, the US and other allied troops set about the task of rebuilding the bombed out remains of the country. Officers were trained to act as interim burgomasters (town mayors) in some of the hamlets that had been damaged and deserted, and Reeves volunteered to be a truck driver, transporting equipment, supplies and people. He just didn't realize that some of the 'people' would be German POWs.

STANDING IN FRONT OF A COMMAND CAR

He was sent down to Manheim, Germany and it was there that twenty or so German POWs climbed into the back of his truck as well as the ranking POW officer who rode in the cab with him. He had no security personnel with him, and only the grease gun and the hope that there wouldn't be any trouble. Thankfully, there wasn't.

"They really had nowhere to go," Reeves said. "They probably felt safer being a prisoner of the US than being taken by the Russians. Or worse yet, to escape out into the countryside where the locals might want to exact some payback on them."

The Germans were used as labor in the rebuilding process, and Reeves transported several loads of them up to Belgium while hauling back rubble on the return trip.

On his many forays into Germany, he and other GIs would often be met by German girls wanting to meet an American soldier. The foreign 'relations' kept up for a while until command found out about it, and then the edict came down.

"No frat with the fraus," Reeves chuckles.

The most memorable day in his army career came when he was on a road construction detail. An Army Col. had wanted a second entrance built for one of the camps and Reeves and another GI were putting down base rock for the road. I was on the bulldozer when I was flagged down by an officer who came up and asked me if I was Private Russell Reeves, and if so, did I have a brother who had been lost in the war?

Reeves answered, 'yes,' to both questions, and the officer went on to explain that the government had just passed a law that if a person was serving overseas and had lost a family member in the war, then they could go home at any time – which meant right then.

"I didn't have to try to decide and so I got the information from him, and left my bulldozer right where it was sitting," Reeves recalls.

He returned home and finished out his duty at Fort Knox, Ky. before returning home. After various jobs, he finally settled on a career with the TVA, where he worked for 34 years before retiring.

His wife has passed, but Reeves still counts as his family; 2 children, 2 grandchildren, 5 great-grandchildren, and 2 great-great-grandchildren. Russell Reeves lives in Gallatin, Tn.

KEEPING THE FAITH

CLARENCE D. DEVEREAUX, PETTY OFFICER 3RD. CLASS, U.S. NAVY

Lots of veterans have a rough time in boot camp; the yelling, the marching, the lack of sleep, but Clarence Devereaux had an entirely different experience.

"It was only the first or second day," Devereaux recalls. "And I started feeling really weak while we were on a march. I stopped and told my superior about it, but he didn't believe me and told me to keep marching."

In short order he was vomiting up blood, then they took him seriously and transported him to sick bay. The year was 1955, and Devereaux had been drafted just a few months earlier and ended up here, at the Navel Training Center in San Diego, California.

In sick bay, they made the quick diagnosis that he had an ulcer, even though he didn't smoke or drink, and sent him to the ulcer ward – which was right next to the psyche ward!

"It was terrible," the young man from Tyler, Tx. remembers painfully. "The psyche ward guys were constantly yelling, but even the guys in the ulcer ward with me were problems as well."

Restricted to a bland diet, some of the other patients on the ulcer ward never-the-less found a way to steal a key to the refrigerator and would break into it at night and eat all of the food they weren't supposed to eat – and they would pay for their sins the whole rest of the night.

Devereaux shakes his head. "They were all moaning and groaning and the psyche guys were yelling all night. I had to ask the man upstairs, 'Oh, God, what did I do to deserve this?' But I knew the Lord would show me the way."

After thirty days, Devereaux was released and had to start boot camp all over again – and it wasn't nearly as bad.

Born in 1933 and the youngest of fourteen children, life was far tougher growing up in the segregated south. Without benefit of an education, his parents had to work as farm laborers to keep food on the table and a roof over their heads. Still, his mother was a strong woman of faith and raised them to be honest, hardworking, and respectful, even to some of the local whites who discriminated against them.

Despite these tough odds, Clarence was able to graduate from college (the first in his family) with a Bachelor's Degree in Business Education. He had just barely started to send out his resume' to apply for jobs when his draft notice arrived.

"My mom was disappointed that after earning my college degree I had been drafted, but she told me that 'we loved our country and when called to serve, we had to answer.'"

During boot camp Devereaux was influenced by some of his fellow recruits to go into dentistry, against his instincts.

"I really wanted to see the world and go overseas, but these guys convinced me that I shouldn't do it in case we went to war. So, I chose (or 'struck for') dental tech."

But even though he chose to go into dentistry that still didn't mean that Devereaux never served on a ship.

"It was called the USS Never-Sail," Devereaux laughs. "And it was sitting on concrete right next to Harbor Drive in San Diego."

USS RECRUIT AKA: THE USS "NEVER-SAIL"

Officially known as the USS Recruit, the ship was a 2/3 scale of a destroyer escort that, although it had been commissioned by the Navy, never in its life touched the water.

"It was crazy! We were looking down at the cars going past on the street and we were supposed to drill like we were a hundred miles out at sea!"

After graduating boot camp, he was trained in dental technology and assigned to work at the El Toro

Marine Station, Ca. which he thoroughly enjoyed. During his time there, he got married and he and his wife had the first of their three children.

ON THE USS "NEVER-SAIL"

Devereaux mustered out of the Navy in 1958 as a DT-3 and went on the road as a Gospel singer, an avocation that he truly enjoyed. Before long though, the financial strains of having a wife and a child to support became too great and he returned to Los Angeles where he took a job as a probation officer with Los Angeles County. While he was working as a probation officer, he answered the call of the lord by becoming an ordained pastor.

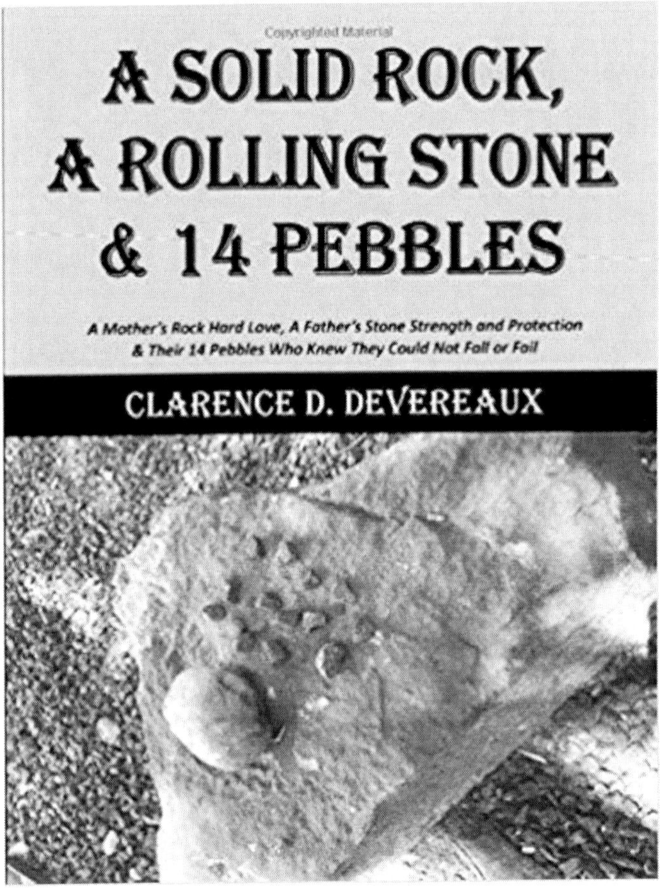

He retired from the county 28 years later and continued ministering for several more years before retiring for good. He and his wife have three children,

4 grandchildren, and 2 great grandchildren. In 2019, Devereaux wrote a book, ***"A Solid Rock, a Rolling Stone, and 14 Pebbles"*** about his life and his faith growing up. The book is available on Amazon.com and it was recently accepted into the Library of Congress.

EXPRESS YOURSELF

SEAMAN 1ST CLASS, U.S. NAVY

DANIEL CAÑAS

Daniel Cańas remembers his mom being very proud of him when she learned he had enlisted in the Unites States Naval Reserve.

"My mom's father fought in the Mexican Revolution along with Pancho Villa," he recalls. "And she idolized him."

Born in 1935 in the newly incorporated city of Gardena, Ca. Cańas described the area as the end of the road, as far as his world was concerned.

"We lived in what you might call a 'Mexican' neighborhood in some tiny houses on Vermont Avenue between 166th. and 168th. The houses were very small, and the paved sidewalks ended at our neighborhood."

After graduating from Gardena High School in 1954, Cańas took a job as a storekeeper with nearby Honeywell Corporation. It was here that he and a friend decided to not let the draft dictate their lives and send them to the Army, but to head off the situation by joining the Navel Reserves. They entered in 1956 and were sent to boot camp in nearby San

Diego. For Cańas, it was a shock to his system, both from a cultural point of view, but also culinary.

"I hated the food," he recalls. "And I didn't eat for two or three days."

He shudders, saying, "What they did to those eggs! Uggh! And that stuff called, 'SOS.'"

But eventually, he got hungry enough and he ate it, even developing a fondness for chipped beef on toast, or SOS.

After graduation several months later, he was assigned to nearby Imperial Beach at the helicopter base there. (The friend he had joined with did in fact get assigned to a ship). But besides missing the experience of going to sea, Cańas also was dismayed to learn that he wasn't going to become a storekeeper, an MOS he had requested congruent with his civilian employment.

"They didn't have any openings," he says. "And so they told me I was going to be doing maintenance work in the dependent trailer park. Essentially, I was mowing grass and watering."

Not exactly living up to the slogan, 'It's not just a job, it's an adventure.'

Still, he performed his duties well and with pride. And soon enough, another opportunity came along and he jumped on it.

"I was to drive what was called the 'crash truck' and work with the Navy Corpsman responding to motor vehicle accidents and helicopter crashes. I also had to work inside with them at our base, working the reception desk and what not."

Cańas enjoyed the duty much better, including the better chow they were provided. He became close to the corpsmen he worked with and they took a special interest in him, even to the point of teaching him how to give people shots. Occasionally, this special skill would lead to some interesting encounters, especially when Cańas had to man the front desk late at night on a weekend.

"We were right next door to Tijuana, Mexico and sometimes the sailors would come in after engaging in some rather questionable behavior," he laughs. "They would plead with me, 'Doc, doc I met this girl and well…you know, I need you to give me a shot.'"

Cańas would just tell them they had to come back in the following day for blood tests, and then send them on their way.

With the Korean War over, Daniel Cańas mustered out of the Navel Reserves after only two years as a Seaman and stayed inactive for four more. He is proud of his service and says that it taught him some very important life skills.

"I learned to express myself better, and when I returned home to my job at Honeywell, I was more confident. I also wanted to achieve higher positions, and I wasn't afraid to let them know."

He eventually rose to the rank of manager of production control at the tech firm, and stayed there until the plant closed. Along the way, he married and had five children. He still lives in Gardena, Ca. to this day, not far from the area where he grew up.

DUTY BOUND

THIERRY WILLIAMS, 1ST. SERGEANT, U.S. ARMY

First Sergeant Thierry Williams stands in front of the classroom at Gardena High School with twenty or so Junior ROTC students focused on him. Dressed sharply in his blue pants with a red 'blood stripe' and a white uniform shirt displaying his many medals, he cuts an impressive figure. The uniform fits in well with his topic today; personal grooming, and the difference it will make in how people judge you. The classroom and the topic are far from the humble beginnings the Louisiana native came from, and from what inspired him to embark on a lifetime of service.

"We were very poor growing up," Williams recalls. "But our home was filled with so much love, we really didn't notice."

Home was Lake Charles, La. a part of the Jim Crow South. Williams was aware of the differences between his life and those of the white kids who lived literally - and figuratively, on the other side of the tracks from him, but he was too busy playing army with his brothers and friends to notice.

"My dad was in the Army Reserves," the First Sergeant explains. "And we would break out his pup tent and set it up in the yard. I also loved playing with little toy army men; that was what got me interested in the military."

And so, after graduating with the final class at his all black high school, W.O. Boston, in 1983, Williams not only started taking classes at McNeese State University, he enlisted in the Louisiana National Guard as well. Before long though, he reasoned that he would make a much better soldier than a student, and so he walked across the street from the college to a recruiting office and enlisted with the United States Army. But it only took the first day of boot camp at Fort Jackson, SC. for that calculation to be severely put to the test.

"I hated it," Williams readily admits. "The yelling, the sadistic DIs, the chaos."

Still, he vowed he would not drop out.

"I had joined the National Guard on the 'Buddy System' with a friend of mine, and I was determined not to wash out and to have to hear it from him."

Boot camp ended mercifully nine weeks later and with it, a new soldier was minted. Williams went on to AIT (Advanced Individual Training) at Fort Lee, Va. and became a Records and Parts Equipment Specialist.

Initially, he supported the motor pool in his mechanized division at Fort Lewis, Wa. and with his skills and work ethic, rose quickly in rank. But then serendipity raised its head and he was reassigned to supporting a Military Intelligence Unit and the first of the UAVs, (Unmanned Aerial Vehicles), or drones, in Florida. He wasn't to stay in the Sunshine State for long though as his expertise supporting drones took him into such far flung places such as Bosnia, Iraq, and K-16 Air Base in Songnam, Korea, where he was promoted to Platoon Sergeant.

By 2007, he had seen enough of the world through the lenses of a GI and retired from the United States Army with the rank of E-7, or First Sergeant. But he wasn't about to stop or even slow down. By now, the man who had considered himself a poor college student had earned his BS in Administration of Justice and was working on an MBA. He transitioned to civilian life and quickly found work as a JROTC instructor at Van Nuys High School.

He taught at the school for only a couple of years before making his way south to Gardena, Ca. where he has been helping to mold young minds since 2011. Since then, the GHS unit has earned the highest honor of the JROTC program in 2015, *The Honor Unit with the Distinguish*. It is represented by a Gold star, which is worn on the cadet's uniform.

WITH GARDENA HIGH SCHOOL JROTC CADETS

Williams credits the military for instilling the discipline, comraderies, and esprit de corps that not only allowed him to succeed, but to thrive as well. More than anything else though, it has instilled in him a sense of purpose to help others, both young and old. He is currently working on his third Master's Degree in Human Behavior, and hopes to work with fellow veterans and the unique challenges they face.

As Williams says, "I just want to make a difference in the world."

SURVIVE, AND SUCCEED

GENE HALE, SPEC-5, U.S. ARMY

NEAR PLEIKU, IN THE CENTRAL HIGHLANDS

He had been at sea for close to three weeks when Gene Hale first saw them. There, from over the horizon, were a small flock of sea birds. Although he had never been on a ship before or sailed across the ocean, he had learned from some sailors on board that the presence of birds meant one thing; you were getting close to land. It also meant that the young man from Birmingham, Al. was a long way from home.

Born one of the younger sons of a hardscrabble family of eleven children, Hale had grown up poor, but proud and honest. His father washed cars for a living, and when Hale and his brothers weren't in school, they had to sometimes get up at 3 o'clock in the morning to help him out.

"We would take the money we made from washing cars and give it to our mom,' Hale recalls. "And she would go to the store to buy food for us for the day. It was truly a hand to mouth existence."

When Hale's older brother was discharged from the Army in California in the mid-sixties, he didn't want to return to Alabama and invited his younger brother to move out west to stay with him. Hoping to carve out a better living than he could at home, Hale jumped at the chance and was soon working menial jobs while at the same time going to college to major in business.

"I had always had an entrepreneurial spirit," Hale explains. "And when I was shining shoes in a barber shop back in Alabama, I learned that I could make additional money by buying the polish from a wholesaler, marking it up, and then selling it to other boys who shined shoes as well."

He was halfway through a two-year program to build on his business acumen when his number came up and he was drafted.

"I really didn't mind being drafted," Hale says. "And looked toward the positive; I would be getting three meals a day, have a steady paycheck, the government would pay my tuition to finish school, plus I could also use the GI bill to buy a house for my mom, a promise I had made to her."

After processing in Fort Polk, La. Hale and the other recruits were sent to Fort Lewis, Wa. for basic training as well as combat training. They were being prepared as hundreds of thousands of other men were in the late sixties; to go fight in Viet Nam. And then one day, it happened.

"We were supposed to be drilling," Hale remembers. "And then they told us to get all of our gear ready – we were shipping out."

After 22 days at sea, their ship arrived at Tuyoa Beach in South Viet Nam. Hale and his fellow soldiers got off and, after seeing the brilliant white sand beaches and bikini clad girls thought that maybe Viet Nam wouldn't be so bad. Paradise would soon be lost though as 4 hours later they were on their way into the central highlands for a different kind of 'action.'

"We got off the trucks and there was no base, no camp – anything, we had to make it ourselves."

'Making it' meant digging foxholes to sleep, fight, and live in. Four days later, they would be in their first firefight as the ad-hoc camp was attacked.

It would be one of many battle experiences Hale endured during his time in combat. But in between the killing and the misery, there were tender moments as well, such as the time Hale became the adoptive father of a young Vietnamese girl.

"She was only about six or seven," he recalls wistfully. "And I don't think she really had anybody in the world, no mother or father to look after her, so I kind of adopted her and tried to take care of her."

Hale would share his C-Rats with her, as well as any candies he came across. He also took one of his jackets, retailored it and gave it to her so she could have some decent clothes to wear. After he left, he often wondered what happened to her.

After doing his obligatory one year in-country, he was sent stateside to muster out of the United States Army in 1971 as a SP-5, Sergeant. As he had planned, he used his GI Bill to not only finish his education, but to buy a house for his remaining family back home. (His mom had passed 2 months after he moved to California to be with his brother).

Although he didn't speak about his experience in the war for over twenty years, Hale was fortunate in the fact that he didn't suffer from PTSD, even though he certainly can understand the affliction and has sympathy for those who have to deal with it.

AT THE COMBAT CAMP IN THE JUNGLE

He credits his coming through on the other side with learning to compartmentalize and move on. He also says that he was fortunate to learn to survive while still maintaining his sensibilities.

And finally, he credits God for keeping him safe so he could go on and do the Lord's work. Besides the numerous citations he received for his service, he has gone on to be a highly successful businessman. He's the CEO of G&C Equipment in

Gardena, Ca. which is one of the largest owned black businesses in the nation. He's also been appointed to numerous councils and commissions by such leaders as Former Presidents: Obama and Bush, Former California Governors: Gray Davis, Arnold Schwarzenegger, and Jerry Brown, as well as many other elected officials. He is currently the chair of The Greater Los Angeles African American Chamber of Commerce, as well as the Chair of the Gardena Police Foundation.

Gene Hale lives with his wife of 37 years in Gardena, Ca. They have two children.

THE SILENT WARRIOR

JAMES YOSHIO MORITA,

CORPORAL, U.S. ARMY

"He was always a very quiet person, about everything." Barry Morita recalls about his late father. "But especially about his time in the war."

Ann Shima (née Morita) nods in agreement with her brother. "There were no pictures of his time in the army around the house, and no mementos."

Then a phone call came from Jeff Morita, Ann's cousin in Hawaii and a retired Army Sergeant and former Gardena, Ca. resident himself. It was like a bolt out of the blue.

"We found out that our dad was in the famous 'Go For Broke' 442," Ann recalls. "This was the unit made up of Japanese Americans that were part of some of the most horrific battles in the war, *and* that he was part of the unit that saved the famous 'Lost Battalion.' We had no idea."

Jeff informed Ann and Barry that because of their father's courageous fighting in France during WWII, he was eligible for the Legion of Honor, the award bestowed on the French Chevaliers, or knights. But they needed to move fast, because James was in poor health physically, and the paperwork had to be submitted while he was alive.

Through Jeff's diligence and hard work a picture began to emerge, a picture of the little known life of James Yoshio Morita, a life that had eluded the family for seven decades.

Born the son of a farmer in the area of the Dominguez Hills, Ca. young James had the distinction of being the first AJA (American of Japanese Ancestry) to be drafted in the Army in Los Angeles County. The date was November 23, 1940, a full year before the United States would enter WWII.

After graduation from boot camp, Private Morita was transferred around through various units: the 29th. Engineers in Oceanside, the 102nd. Radio Intelligence Company, the 54th. Signal Battalion, the Corp Area Service Command #1902, and then to the Detachment Quartermaster Corps in Fort Snelling, Mi. It wasn't until March of 1943, that the now Cpl. Morita was incorporated into the newly formed 442 Regimental Combat Team in Camp Shelby, Ms. as a rifleman in F- Company.

After a year of infantry training, Morita and his unit shipped out to the European Theater of Operations. From June 9th. 1944, they fought on Italian soil. The Italians reached an armistice with the allies on Sept. 3 and by Sept. 5th. Morita and the 442 were on their way again, this time, to liberate France.

Positioned around the town of Bruyeres in eastern France, Morita saw fierce battle and house to house fighting until the town was liberated on October 18th. To this day, the town holds an annual celebration to remember their brave saviors.

But victory or not, there would be no rest for the battled hardened and triumphant 442 as they next moved on to the area around Belmont-Biffontaine. It was here, from October 19th through the 24th that they fought to rescue the famous, 'Lost (Texas) Battalion.' Morita was assigned to the O'Conner Task Force, a unit that would fight and hold off the Germans while the other units rescued the Lost Battalion. Again the fighting was fierce and the Germans employed 'Tree Burst' tactics with their rockets and artillery in the forested hill, causing huge limbs to come crashing down like swords onto the allied forces below.

The Lost Battalion was eventually rescued, but at a huge cost. Altogether, the 442 saved 211 men, but at the expense of 800 casualties, including 121 dead. For their actions, the members of the 442 would later be bestowed the distinction of honorary Texans.

After only a couple of days rest after their victorious rescue of the Lost Battalion, the scant remaining members of the 442 were again ordered back into combat by General John E. Dahlquist, a decision that was considered reckless and incompetent then, *and today*.

It was here on November 2 1944, while continuing to fight against the Germans in the Forest Dominiale du Champ, that Morita was struck by a piece of shrapnel and collapsed into a ditch by the side of the road. Thinking he was dead, the other soldiers left him where he lay and he spent a freezing cold night until the following day when soldiers assigned to recover the dead found him alive.

After he was evacuated and treated for life threatening injuries in several European hospitals, he returned to the US. With the war in Europe quickly winding down, he was honorably discharged on December 5th 1945.

SOME OF CPL. MORITA'S MANY MEDALS

James Yoshio Morita died on April 3rd. 2015 and so reticent was he to discuss his military service that mourners at his funeral were surprised to see a military honor guard and the US flag presented to the family, remarking to themselves and others, "I never knew."

THE CONGRESSIONAL GOLD MEDAL

Cpl. Morita's citations include: Recipient of the Congressional Gold Medal, the Bronze Star, the Purple Heart, the Distinguished Unit Badge, and many more. On January 14th 2016 he was posthumously award the French National Order of the Legion of Honor.

THE FRENCH LEGION OF HONOR

GROWING UP & STANDING UP

LEE HULL, AIRMAN, U.S. AIR FORCE

Airman Lee Hull (née Beach) marched smartly into the Sergeant's office to speak her mind. For several weeks she had been in charge of a work detail at Carswell Air Force Base in Fort Worth, Tx. and had grown weary of the machinations of her superior trying to inflict his discreet form of racism on her group, and on one African American Airman in particular. For Hull, who had spent her formative years growing up in the more diverse and tolerant Hawthorne, Ca. being assigned to the Texas base had been her first real experience with prejudice and bigotry.

"I know what you're trying to do and you have to lay off of her," Hull told him in no uncertain terms. "She's a good worker."

Although it was a risky move, confronting the Sergeant did the trick and he apologized and they were never bothered by him again. The incident wasn't the first or only time that Lee had stood up for herself and for her group, but it was something she could scarcely imagine herself doing when she enlisted in the Air Force three years earlier.

"As a young girl, I was very shy and unassertive" Hull recalls. "But the Air Force really taught me that I had to learn to stand up for myself and speak up."

Her family had moved from Kansas City, Ks. in 1951when Lee was about 13 years old. She attended the local schools and graduated from Leuzinger High School in Lawndale, Ca. in 1956. She wanted to continue her education and go on to college, but her parents didn't have the money and so she was on her own to try to raise it. It was here that her shyness and timidity did her the most damage.

"I needed to get a job to earn my way through college, but when I would walk into a store they would say, 'we're not hiring' and I would just say, 'okay' and walk out."

The military and GI Bill provided her another avenue to finance here higher education and so in 1956, she enlisted in the United States Air Force.

Basic training was at Lackland Air Force Base in San Antonio, Tx. and it was here that she not only overcame her lack of assertiveness, she also came to grips with another fear.

"We had to pull guard duty at night," Lee says. "And I was terribly afraid of the dark. But we had to go downstairs in the dark and check the casement windows. I was terrified, but I knew I had to do it. And so I did, and I was very proud of myself and was never afraid again."

After basic, she was sent up north to F. E. Warren Air Force Base to learn her assigned job as a teletype operator. She did well and was subsequently sent back down to the Lone Star State and Carswell Air Force Base where she spent the next several years reading the cryptic holes and dots of the teletype tape. She performed so well that she was even assigned to a select group that would have to set up a mobile communications network if the country ever went to war.

"We drilled in the emergency procedures, but we always seemed to know when they were going to test us, so we weren't really surprised when they happened and knew it wasn't the real thing – except for one time."

AIRMAN HULL WORKING AT HER TELETYPE MACHINE

An alarm sounded unexpectedly one night and Hull and her roommate - who was also part of the group – looked at each other thinking, 'Oh boy, this is the real thing.'

"We each grabbed our gear and started heading out the door. We were about halfway there when all of a sudden it came back that it was a false alarm."

As it turns out, a plane experiencing electrical problems was not able to identify itself and was entering California air space, triggering an unidentified aircraft alert.

Hull left the Air Force in 1959 and quickly found work in communications for Pan Am Airlines. She worked for the carrier for thirty-two years before it went bankrupt in 1991. By then, she had already married and moved to Gardena, Ca. with her husband and three children. She worked for United Airlines for another ten years before retiring permanently.

Although she didn't make a career of the military, she values her time serving and how it allowed her to grow. She also thinks it could do the same for others.

As Airman Second Class Lee Hull puts it,

"I think everybody, girls and women – everyone, should do at least one year of military service."

NO REGRETS

JOHN P. KAPER, CORPORAL, U.S. MARINE CORP

The AK-47 round hit John Kaper like the proverbial ton of bricks, penetrating in his groin area and travelling along the femoral artery of his left leg, destroying the vital artery in the process. It was what was considered to be a mortal wound. Already, Kaper, acting as squad leader for his unit in Charlie Company, 1st. Battalion, 1st. Marines, had watched a fellow marine die in the firefight that raged around him and he knew one thing at that moment; he didn't want to be the next.

"I had God, my point man, and the corpsman all on my side that day," Kaper recalls vividly.

After being struck by the rifle round, Kaper instinctively did what many others did in the moment; he jumped up, exactly the wrong thing to do in a firefight. Luckily for him, his point man pulled him down and held him until the corpsman could get to him.

"I was lucky enough to be wearing a web belt around my trousers that day," Kaper explained. "Most guys didn't. And the corpsman took it off of me and used it as a tourniquet on my leg."

It was one of the few things that went right that day, a day that Kaper had spent years trying to arrive at.

After dropping out of Santa Monica College in 1966 where he had been playing football, John Kaper signed up to do what he had always dreamed of: being a Marine. He had several friends that were marines, and he knew that he wanted to be a part of the best and the toughest fighting units in the world. He also knew that he wanted to be in the thick of things.

"Viet Nam was already ramping up, and I knew that I wanted to go and to fight there. So I requested infantry and volunteered for Viet Nam. I didn't want to do anything else."

But 'anything else' was exactly what he got.

After breezing through boot camp in San Diego, Kaper was disappointed to learn that because of his brief – albeit valuable, college education, his designated MOS was 0141, an office clerk. Wish #1, to be a Marine, had been fulfilled, now he had to start working on the other two.

He was sent to Cherry Point, NC. in July of 1966 and was tasked with handling special court martial cases. He did this for several months before being able to volunteer for Viet Nam and being sent back to Camp Pendleton before shipping out along with 1,500 other Marines and 3,500 Army troops on a transport ship that would take them all the way to Da Nang, arriving in April of 1967. Wish #2: *check.*

In Viet Nam, Kaper was assigned to the 7th. Armored Transport Company Alpha, and moved to Chu Lai, a Marine Corp Military base and the second military airfield to be constructed in the country. The transport unit was responsible for running convoys to move Marines and supplies to the units in the field. Kaper was still being utilized as an office clerk, but he did go out on the convoys and would volunteer to man the .50 caliber machine gun that was mounted on some of the vehicles. At this point, he was getting closer to the action, but still shy of achieving his third wish.

After a couple of months, he was sent back to Da Nang which was a hotter location and closer still to the action. It was at this point that being a clerk paid off for Kaper and allowed him to achieve his third wish.

"As a clerk, I was privy to reading all kinds of correspondence," Kaper explains. "And this meant I had an inside track to what was going on in the field with troop strength and so forth."

He finally saw his opening and jumped on it. In October of 1967, he was officially transferred to infantry. Wish #3: COMPLETE!

He was sent to the Dong Ha Base and then to Con Thein and was assigned to Charlie Company, 1st. Battalion, 1st Marines. Within a week he was already tasting combat when Charlie Co. was asked to help extract Alpha Co. when they got pinned down by the enemy. The action continued unabated and after cycling through Quang Tri, he was sent home for thirty days leave in February of 1968, just missing the Tet Offensive.

IN QUANG TRI

A month later he was back 'in country' and was sent to Hill 881, just 3 miles from the Laotian border and the Ho Chi Min Trail. Three months later on June 1st, Charlie Co. was ordered to sweep the hills outside of Khe Sanh. It was here, just four days later on June 5th, that Cpl. John Kaper would nearly meet his maker.

"I was losing a lot of blood and fading fast," he recalls after sustaining the gunshot wound. "I was also hoping that the tourniquet wouldn't slip off as they transported me to the medivac helicopter down below."

Luckily it didn't, and after a 15 minutes flight to the hospital in Khe Sanh, the doctors were not only able to save his life, but his leg as well.

He received his purple heart aboard a hospital ship and was sent home for 8 months of rehab at the Long Beach Naval Hospital.

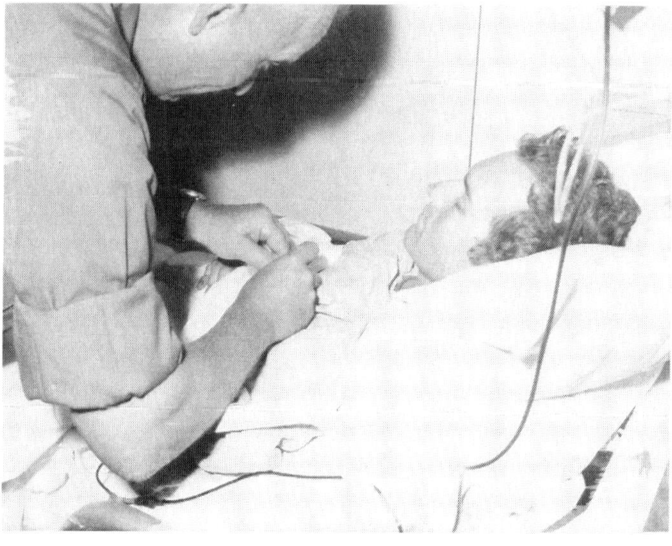

RECEIVING THE PURPLE HEART IN HIS HOSPITAL BED

He mustered out of the Marines in March of 1969, and although he suffers from the effects of his injury to this day, John Kaper has no regrets about enlisting and serving his country. He lives in

Gardena, Ca. with his wife Pat, and still gets together with his unit for reunions.

SEMPER FI!

A HIGHER CALLING

MARK HENDERSON, PETTY OFFICER 3RD CLASS, U.S. NAVY

One of the most popular recruiting slogans of the US Navy was, "It's not just a job, it's an adventure!" But Mark Henderson had a different goal in mind when he raised his hand and swore in to become a part of the largest and most powerful maritime force in the world.

"I was attending school at Los Angeles Southwest College and pursuing a liberal arts degree," the 22-year Gardena, Ca. resident recalls. "But I wasn't really sure how I was going to continue to fund my education. Then I heard about the GI bill."

His intention was to join up, do his four years, and then his pursuit of a business degree, but then life happened – more specifically, military life.

"I was trained in electronics to become a sonar technician, and I found I really liked it."

Following eight weeks of basic training in San Diego, he was sent to electronics training, one of the toughest schools in the Navy with a staggering 80% wash rate. Henderson excelled at it and was assigned to the USS Fife, DD-991 Spruance Class Destroyer as a member of the Midway Battlegroup, stationed in Yokosuka, Japan.

He was part of the ASW (Anti-Submarine Warfare) group whose main mission was to hunt enemy subs around the world. Besides maintaining the acoustic equipment designed to listen for the underwater predators, Henderson also did 'low frequency gram analysis,' which enabled him to determine the type of vessel, its heading and speed.

He found that he loved his job, his fellow seaman, and especially the feeling of oneness, that part of being something more important than you. Pretty soon, the GI Bill money didn't seem as important to him. This was in spite of the fact that he suffered with terrible sea-sickness.

"It became part of a joke," Henderson laughs. "Every time I had to stand watch when we were at sea – which was most of the time - my shipmates would say, 'There goes Henderson again with his box of crackers.'"

Then one day, after about a year and a half into his service, he decided he was tired of it and refused to feel sick – and that was that.

Nausea was the last thing on his mind though when their ship was diverted to Chinhae, South Korea in late 1990. Saddam Hussein had just invaded Kuwait the previous summer. Sabers had been rattled, and tensions raised throughout the military world. Finally, the situation came to a head and Henderson and his shipmates had to load 52 Tomahawks missiles onto the Fife, before continuing on to the Persian Gulf. It was, he remembers painfully, one of the toughest times during his four and a half years in the service.

"We had to load these things up wearing full 'Chem' (chemical) suits, which are made out of Mylar and incredibly hot. On top of that, we had to wear our full uniforms underneath!"

Getting on station in the Persian Gulf was another experience onto itself. It was an ominous feeling, and he remembers being scared, not necessarily for his own safety, but for the overall gravity of the escalating situation.

"I remember seeing the oil wells on fire, and the mines in the water, and I realized then that this was the real thing and not a drill; we were going to war."

But there were also lighter times. In June of 1991, the volcano Mount Pinatubo erupted in the Philippines, covering much of the country in a thick layer of ash. Henderson and his shipmates were part of a humanitarian effort to clean up the ash from some of the affected areas.

"It was a very thick layer of ash," he recalls. "And it had rained the night before, which made it stick together like snowballs. And it only takes one guy…"

And that's when it started.

Before long, in what could only be described as a 'boys will be boys' moment, the work detail quickly devolved into a full blown 'ash-ball' fight.

Henderson laughs fondly at the episode, and the aftermath.

"And then of course, nobody was thinking far enough ahead to realize that we had to return to the ship and our CO would like an explanation as to why we were covered in ash!"

Besides the ship-based sonar systems, Henderson also became involved in the airborne ASW exercises, flying in a helicopter and dropping sonar buoys to track enemy subs.

He mustered out of the Navy after 4 ½ years at the rank of Petty Officer Third Class after earning many commendations, including the Meritorious Unit Commendation Medal for valorous or meritorious achievement or service in combat. He continued his higher education and earned a Bachelor of Science degree in Telecommunications Management from DeVry University, a Master of Business Administration degree from the University of Redlands, and a Doctorate in Organizational Leadership from the University of La Verne. Most notably, he remained in public service as the Information Technology Manager at Pierce College.

Mark Henderson was also elected to the City Council of Gardena, Ca. in 2015. He remains active in his community and numerous organizations including PV Jobs Inc., which works to train community members and Veterans in obtaining employment within the Building Trades.

After spending his entire adult life in public service, first in the military and then working in the public sector, he says about public service, "One of the things I like most about being in public office is being responsible to others."

Truly, a higher calling.

THE ARTIST'S WAY

MICHAEL BRENNAN, AIRMAN 1ST. CLASS, U.S. AIR FORCE

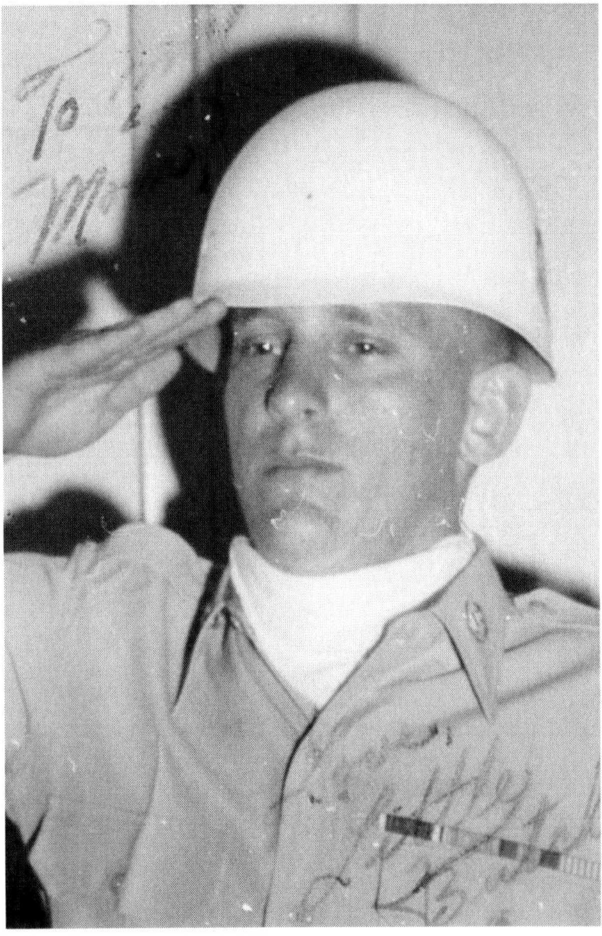

To this day, Michael Brennan still considers the entree he ate that fateful day to be the best meal of his life. After swearing in with the United States Air Force in downtown Los Angeles, he was given a 'chit' to get a meal at a local diner before he headed out on the bus to Texas, and he ordered meatloaf.

"It was the best meal I ever had," the Gardena, Ca. resident recalls. "It was empowering because it was the first time I was on my own, and I knew that if I failed, it was on me."

To this day he still considers it his 'victory meal' when something good happens.

But up until this point in his life, it had been a long time since anything good had happened. After his biological father died during the "Battle of the Bulge" in 1944, he had been given up for adoption in his birthplace of Chicago, Il.

"We lived on the south side, and it was getting very rough," Brennan remembers. "And when a relative of ours was killed, my adoptive parents packed up and we headed west, to California.

He attended Saint Anthony's and then moved on to Juniper Serra High School, where admittedly, he was a bad student.

"Academically I was fine," he says. "But my attitude was very bad and I didn't feel like I was going anywhere. And so, after only one year of high school, I told my parents that I wanted to join the military, and they agreed."

Even with their blessing though, he still had a big problem; at sixteen years of age, he was too young by one year. Undeterred, he pressed on with his own unique plan.

"I did the artwork to forge my baptismal certificate, and I think I did a pretty good job," Brennan laughs.

The military was a wise option, and he said that his attitude improved remarkably. After basic training at Lackland Air Force Base, he was sent to tech school to study to be an inflight radio operator on the B-36 bomber. It was to the young man, his dream job.

"It was the most marvelous plane I had ever seen," he recalls fondly. "With six pusher props and four jet engines."

Life was good until the military decide to scrap the B-36 program for various reasons, and suddenly, Brennan and his fellow airmen were out of a job. They were each given new assignments, which was fine, except his new assignment in North Viet Nam would require a higher security clearance, and that's where they caught him.

"They found out I was only sixteen and told me I had to go home. But they liked me and told me to return when I was seventeen, which I did."

Upon his return, he tested for work with computers and did very well. So well in fact, he was assigned to NORAD in Colorado Springs, Co. The year was 1959, at the height the cold war, and one of the most dangerous times for America. Brennan worked with both the IBM 705 and the 1401, at the time, the most advanced computers in the world. He was also a member of the honor guard.

He enjoyed his job at NORAD, but it wasn't without its stressful times. In October of 1962, with tensions at their highest point between the US and the Soviet Union, he was locked down during the Cuban Missile crisis.

"I remember being told to get my weapon and go to the computer room. I also had to call my wife and say goodbye to her. We thought it was the end."

Along with the rest of the country and the world, he survived the crisis and was even visited by President Kennedy shortly thereafter.

"We were all told to remove the firing pins from our weapons," Brennan said. "And when he shook my hand, I was so enthralled by the young president, I couldn't even look at him."

"I remember his shoes," he laughs.

During his time at NORAD, he maintained his love for art and continued to produce works. Then, in the early sixties, he met Sargent Rodriguez who was the art editor of Air Force Times Magazine. On the side Sgt. Rodriguez did portraits of military veterans including many for the Pentagon.

"I was so impressed by him and inspired by his work, I decided that at some point, I wanted to do veteran portraits myself."

Brennan estimates that he's done approximately 40-50 portraits so far.

BRENNAN TODAY HOLDING A SELF-PORTRAIT

He mustered out of the Air Force in 1964 and returned to California where he worked at several jobs including Plastiglide and the Todd Ship Yards, before eventually settling on teaching art. He taught at various locations including LAUSD and at Torrance Unified, where he taught for 31 years.

But the military continued to be a siren song for him and in the 1980's he joined the Army National Guard where he stayed for several years before mustering out for the final time after suffering from complications due to a broken ankle.

Brennan still paints and will do portraits of military veterans at no cost. If you would like to contact him regarding a portrait, his email is: JMBrennan@dslextreme.com.

TALES FROM THE 'WE' GENERATION

RICHARD CASTILLA, SEAMAN, U.S. NAVY

Because they were on the other side of the international dateline when it occurred, VE - Day, or Victory in Europe Day, occurred not on May 8th as is recorded in the history books, but on May 9th for Seaman Richard Castilla and his shipmates in the South Pacific.

"Every year on that date, I pour myself a drink and have a toast." He says.

But the annual libation isn't to celebrate the ending of the horrible war in Europe, it is to honor the memory of his shipmates that were lost on that tragic evening; because that was the night their ship was hit by a Kamikaze, and thirty-seven sailors lost their lives.

Growing up in South Los Angeles, Castilla was only seventeen years old when he did what many others his age were doing; he signed up to fight in World War II. It was August 16th, 1943, and he had just graduated from J H Francis Polytechnic High School a few months earlier.

"My brother had joined the Army a year or so earlier and was in the glider troops," he recalls. "But most of the kids in the neighborhood kids were joining the Navy, including a buddy of mine, so we joined up together."

It was not uncommon as Castilla remembers that times were different back then, especially people and their sense of duty to their country.

"Everyone was patriotic back then. I know that we are referred to as the 'Greatest Generation,' well I think we were the *'We Generation.'* And unfortunately, today, we seem to have the *"Me Generation!"*

After basic training in San Diego, Castilla was trained as a Fire Control Optical Range Finder and was responsible for determining the range to a target using an optical range finding device. His station was up on the flying bridge, the highest point on a ship.

That ship would become the USS England DE 635. Only 306 feet long with a slender 34 foot beam, the newly constructed destroyer escort had a crew of about 215 sailors. It also had a rather somber history when it came to its namesake.

"It was named after Ensign John C. England, who was on the Oklahoma when it was attacked in Pearl Harbor," Castilla recites proudly. "England dove back into the ship three times and saved the lives of three of his fellow sailors, but on his fourth attempt, he never came back."

USS ENGLAND

Home ported in San Francisco, Castilla and his shipmates performed sea trials for several weeks up and down the coast of California until finally they were ready to enter combat. They shipped out on Valentine's Day, 1944. It was a day Castilla will never forget, along with what the first few days at sea were like.

"I was seasick for about three or four days," he recalls painfully. "And I carried a five gallon bucket wherever I went to throw up into."

Seasickness was a common malady for most of the newly minted sailors, but it had its comical moments as well.

Castilla laughs.

"So many guys were sick on the ship that they were puking into the trash cans. The cans smelled terrible but worse yet, when we hit rough seas, they would tip over and the floors would be covered in puke!"

They made a brief stop in Pearl Harbor, which Castilla remembers as still badly damaged.

"It was very emotional, seeing all of the destruction," he says.

After Pearl Harbor, they continued on to their theater of operation near Guadalcanal where they engaged in 'picket duty' to locate and destroy enemy subs. On May 13th, they broke a Japanese code and the hunt was on to kill as many subs as they could.

And boy, did they!

On May 19th, 1944 the England detected a Japanese sub designated I-16. They unleashed their salvos of Hedgehog Anti-Submarine Mortars and after a few misses, finally scored a hit with the fifth attempt. Four to six detonations of the Hedgehogs on the hull of the sub were followed by a large underwater explosion which was so powerful it lifted the England's fantail and knocked men off their feet. Debris began floating to the surface twenty minutes later. The following day, a three by six-mile oil slick marked the location on the calm surface of the Pacific.

But the crew of the England were just getting started.

"We sunk six Japanese subs in just twelve days," Castilla says proudly. "It's a record that's still standing to this day, and our ship received a Presidential Unit Citation."

There was no time to celebrate though as the final push toward Japan and Allied victory were in the crosshairs. They continued their patrols and working their way toward their eventual target.

CASTILLA (FAR LEFT) WITH HIS CREWMATES ON THE ENGLAND

Slightly less than a year went by and Castilla and the crew of the England found themselves at the US Navy's secret base at Ulithi, 1,300 miles south of Japan. Hundreds of other ships were docked there and staging for the coming siege of Okinawa. Then, on April 1st, 1945 one of the bloodiest battles in the Pacific was launched.

It was early evening on May 9th, while supporting the battle for Okinawa that the England's radar spotted three Japanese Aichi D3A's dive-bombers (also known as 'Vals') heading inbound to attack them.

Luckily for Castilla and his crewmates, two Marine planes were patrolling nearby and were able to shoot down two of the enemy aircraft. But a third one, a Kamikaze, got through and hit the ship directly under Castilla's station on the flying bridge.

"It was an incredible explosion," he remembers, "And devastating. Twenty Seven crewmen were killed instantly, and ten more were blown off the ship into the water, never to be found again."

Fire from the explosive and fuel laden plane raced up the side of the structure towards Castilla and some fellow seamen. They had to climb out of the flying bridge and come down the other side in order to stay alive. Then they scrambled back to the other side of the ship to assist with the firefighting effort, which took several hours.

KAMIKAZE DAMAGE ON THE USS ENGLAND

The England survived, but was badly damaged and had to return stateside for repairs. By way of San Diego, they made their way down the coast and went through the Panama Canal, which Castilla says was very interesting.

"We always had to travel with no lights at night because of the war in the Pacific. But when we went through the canal and were on the Atlantic side, we could finally turn on our lights, because the war in Europe was over."

THE ONLY REMAINING PIECE OF THE ENGLAND. IT WAS LOCATED JUST UNDER THE FLYING BRIDGE WHERE CASTILLA SERVED. IT'S CURRENTLY DISPLAYED AT THE ANNAPOLIS NAVEL ACADEMY

After docking at the shipyards in Philadelphia the crew was given some much needed leave. It was during his time home Castilla that he heard that the Japanese surrendered. World War II was finally over.

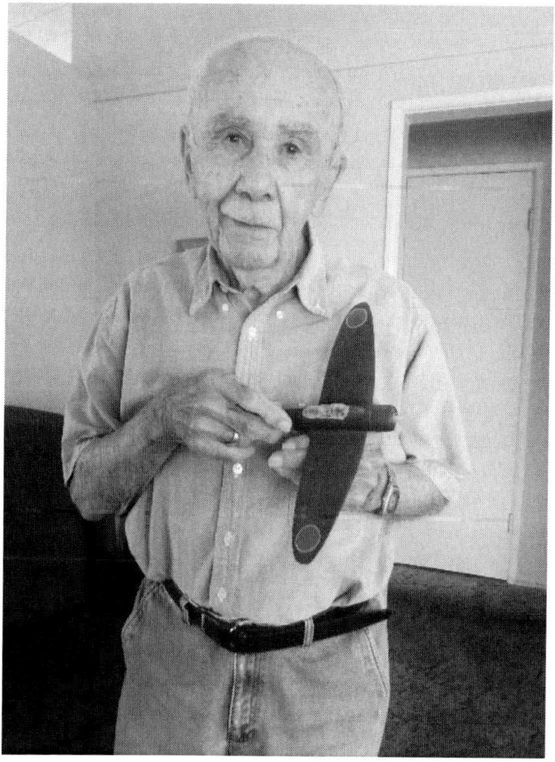

SEAMAN CASTILLA TODAY, HOLDING A MODEL OF THE VAL, THE KAMIKAZE PLANE THAT HIT HIS SHIP ON VE-DAY

Castilla discharged out of the Navy in 1946 and took a job at General Motors in Southgate, Ca. He rose quickly in the ranks and in 1957 purchased a home in Gardena, Ca. where he still lives to this day. Along the way, he married and had 4 children, 8 grandchildren, and 5 great-grandchildren.

A proud member of the "We Generation.'

NUMBER THREE

RON JACKSON, CORPORAL, U.S. MARINE CORP

Ron Jackson quickly learned the order of priority that the North Viet Namese placed on its targets during an ambush. First was the person in charge. Second, was the radio-man. And third was the machine gunner – his MOS. Still, he'd rather take his chances here in the jungles of Viet Nam than dealing with his life back home. Here, he had a gun, and a big one at that.

"The other guys in my unit may have not liked me," Jackson recalls. "But they certainly respected me and had to rely on me when the stuff hit the fan."

It was 1968 and the Shreveport, La. native was busy patrolling as part of his unit's counter-insurgency efforts in the jungles of Viet Nam. After a troubled childhood that saw him doing stints in reform schools after moving to California to live with his sister, he returned to Louisiana in his senior year to graduate high school.

"I didn't want to have my high school diploma come from a reform school," he admits.

After graduating from Booker T Washington High School in his home town, he saw little if no opportunities that awaited him and so he enlisted with the United States Marine Corp.

Bootcamp was in Camp Lejeune, NC. and to the young Jackson, who had been toughened on the street, it was not a big deal and he adjusted easily. Still, he witnessed firsthand the creative forms of punishment the DI's could dish out.

"There was one guy," Jackson laughs. "Who thought he could get away with smoking a cigarette when he wasn't supposed to. The DI put him under a poncho and lit the entire pack for him to have to inhale. He never did it again."

After graduating Jackson did his leave and then returned to Camp Lejeune for advanced training. As an 0331 infantryman, he was trained on several weapons: the .45 pistol, the M-14 rifle and the M-60 machine gun, which he took to handily, scoring as an expert. From that point on he would carry the weapon on patrols in Viet Nam, making him a juicy target.

He arrived in-country in February of 1968 and was assigned to a CAG or Combined Action Group as an advisor and a trainer. His small unit occupied a compound across from a small village where they kept an eye on the locals as well as trained the South Viet Namese soldiers. It was here that he experienced targeted racism and bigotry far from home.

"I was the only black guy in a unit of white guys," Jackson says. "And I got messed with a lot. And not just by my own guys, but by the South Viet Namese as well. They picked up some bad American habits, such as using the 'N-word' when referring to me."

Still Jackson had a job to do and knew that in order for him and his unit to survive, he had to do what he had to do. On patrols, when they were caught in an ambush, he was the go-to guy when he lit up the big M-60 with its long belts of high powered ammunition.

In one case however, they were attacked by insurgents in such close quarters while out on patrol that they couldn't use their guns, and had to fight the enemy hand to hand.

PUMPING IRON IN OUR COMPOUND IN VIET NAM

Jackson can't recall how many patrols he went on in Viet Nam, but he carries a reminder of one in his body to this day, though he's not sure what it is.

"We were out on patrol and the guy next to me stepped on a booby trap and was blown up. I caught a piece of shrapnel in my leg. The doctors couldn't remove it so it's still inside of me, although I don't know whether it's metal, a rock, or maybe even a piece of bone from the guy who died."

He returned stateside in March of 1969 and finished his duty as a brig guard at Camp Lejeune before mustering out of the Marines on July 1st, 1970 as an E-4, Corporal. Besides his Purple Heart, he was awarded numerous other citations.

Ron Jackson lives in Gardena, Ca. with his family and is supportive of his fellow veterans. He once provided the photo documentation of the construction of an award winning float in the 2013 Rose Parade, commemorating the 50th. anniversary of the Korean War.

STILL GOING...

NOBORU KAGAWA, TECHNICAL SERGEANT, U.S. ARMY

Tech Sergeant Noboru Kagawa was stunned when the medic who was treating him for the flu told him that his (The doctor's) brother was dating Noboru's old girlfriend back in California.

"Oh, my brother is going around with her," the doctor said "And he's really in love with this girl."

The place was WWII Italy, the year 1945. And it was at that moment that Kagawa realized that the world was indeed a small place!

Home had originally been in Gardena, Ca. Born in 1924, Noboru, a Nisei, or second generation Japanese-American, and his family farmed in the Gardena Valley, raising cauliflower, celery, and young berries.

Life was good if simple; the US had not yet entered World War II, but it wouldn't be long before they did. It was news that Noboru and his brother heard with a resounding bang.

"I was in a truck with my brother driving home after delivering a load of produce when somebody hit us from behind," Noboru recalls vividly. "I didn't know what the heck was going on because at that time we didn't even know that Japan attacked Pearl Harbor. And we looked at the guy that ran into us and all of a sudden he yelled, 'You goddamn Japs!' And I thought, gee, what did we do? Then I turned the radio on and we found out what happened."

It wasn't long afterward that the US Government came in and told them that the land the Kagawa's leased on 190th. St. between Western and Normandie had to be used for defense purposes, and then they confiscated it from them.

His family wasn't sent to an internment camp like many others, but they had to evacuate. They drifted north to Rexburg, Id. and took to doing what they knew best; farming. It was when they were living here that Noboru received his draft notice. The year was 1944.

Leaving behind his family and his wife, Shizue, he was inducted into the US Army on May, 29, 1944 at Ft. Douglas, Ut. one of the oldest military installations in the US and rumored to be haunted.

Noboru wasn't around long enough to make the acquaintance of any poltergeist as he was soon on a bus to Camp Shelby, Ms. It was here at Shelby that Noboru reconnected with his younger brother who was already a part of the famous 442/100 Infantry Regiment, a fierce fighting unit made up almost entirely of Japanese/Americans. The unit's motto was, "Go for Broke!"

After basic and advanced training at Shelby, the 442nd was ready to ship out overseas to Europe in June of 1944. But at the last minute, Noboru found out he wouldn't be going.

"The camp medic, Joe Saito, told me, 'Nob, I'm gonna hold you back. You have a high arch in your foot, and I don't think you going to be able to march in Italy and France.'"

Noboru would join the 442nd six months later in January of 1945, but medical issues would delay him from joining his unit once again; he developed frostbite and trench foot while having to ride in an open truck in Belgium in zero degree weather. It took a month of convalescence in Cannes, France before he was finally able to join the 442nd as they waged a battle in the Po Valley, one of the last major campaigns to drive the Germans out of Italy.

Most of the hills in the area were named after places in the United States, and Noboru and his unit took the one named Ohio first. It was here that he got his first real taste of battle.

“We were pinned down, we could see grenades coming over, and we're throwing our grenades right back at them,” Noboru says. “It was real back and forth, but we prevailed and moved on.”

Before long, Noboru and the 442 had the Germans on the move and retreating. He remembers that they were seeing less and less resistance the farther they pushed and the more the Germans retreated, until finally, the Germans surrendered in May of 1945.

READING THE GOOD NEWS

It was then that the Noboru's unit switched gears from fighting the Germans, to guarding them as prisoners of war in Livorno (Leghorn) Italy. It was an experience that had a profound effect on him.

IN LEGHORN, ITALY

"You know, they're just like you and me," he admits. "There was no hostility or anything because I think they were told to do 'this and that,' just like we were told to do 'this and that.'"

Noboru Kagawa mustered out of the army as a Tech Sergeant on July 18th, 1946. His Awards include: The Distinguished Service Badge, Bronze Star, European African Middle Eastern Theater, WWII Victory Ribbon, Army of Occupation (German), and the Purple Heart (for his frostbite). On November 2, 2011, the Congressional Gold Medal, the highest civilian honor bestowed by the United States Congress was presented to the 442/100. Noboru's granddaughter, Kim Banducci, represented him at the ceremony in Washington, DC.

Author's note: TSgt Noboru Kagawa passed away in 2020 at the age of 96.

FORTUNATE SON

GREG BRELAND, PETTY OFFICER 3RD. CLASS, U.S. NAVY

The flight deck of an aircraft carrier has the ominous distinction of being called *"the most dangerous three acres in the world,"* but Greg Breland didn't care.

"Watching aircraft take off and land on the deck was probably one of the most incredible things I've ever seen in my life," he recalls fondly.

It was a sight he never expected to see, let alone have a front-row seat to.

Born the son of an aircraft mechanic for the Air Force in Las Vegas, Nv. in 1963, his family moved around a lot as his father was transferred from one duty base to another. They finally settled in Jupiter, Fl. where Breland spent most of his years growing up, fishing and having a love for the ocean. Finally, as he was nearing his 18th birthday, he knew he had a decision to make regarding his path in life.

"I always knew that I wanted to join the military, but the Navy wasn't my first choice. I wanted to be a Marine!"

Unfortunately for Breland, or, as he would say now – 'fortunately,' the Marines weren't an option at that point in time as it would take him over a year to get in. He settled on the Navy, and never regretted it.

"Bootcamp was really easy," he recalls, and he made it through with no issues. He even took advantage of an opportunity to join the Navy Chorus, although he had no musical ability.

"They were looking for volunteers to sing, and I knew that it would allow me to get off of base and have other privileges, so I raised my hand." He laughs about it today. " And, they never caught on to the fact that I couldn't carry a tune!"

And even though he had always been mechanically inclined, he didn't sign up to be an aircraft mechanic like his father. Instead, he chose a more artistic line of work.

"I always liked photography, and so I signed up to be a fleet photographer."

But the chemicals used in the development of film didn't agree with him, and he quickly broke out with skin rashes. He was released from the program and would have to begin training for his alternate

Military Operational Specialty (MOS): aircraft mechanic.

Unfortunately, there was a year or so before another AMS (aircraft mechanical school) was set to begin and he became what the military terms, a 'non-rate.' He was sent to the fleet and to the USS John F Kennedy aircraft carrier where he became an 'aircraft fueler.' It wasn't the most glamorous job, but it got him close to the planes – and to the action…which he loved!

Besides the thrill of being on the flight deck and seeing the planes take-off and land, his year of lag time with fleet flight operations came in handy when he was finally sent to Millington, Tn. to attend AMS a year or so later.

"Being the most senior person in there, I was made the Petty Officer of the class."

And if Navy bootcamp was easy, the basic AMS school was almost laughable.

"The instructor had a bit of a code," he chuckles. "If a question was going to be on the test, he would stomp his foot three times on the floor to let us know."

It would be the last easy time he was to have.

After graduating from AMS school, he went on leave to visit his extended family, in Kingsport, Tn. before moving down to Jacksonville, Fl. for specialty training on the plane that he would work on for his remaining time in the Navy: the Lockheed P-3 Orion.

P-3 ORION

First seeing service in 1961, the P-3 Orion is a four-engine, turboprop anti-submarine and maritime surveillance aircraft developed for the United States

Navy. The P-3 had the ability to locate submarines using a Magnetic Anomaly Detector (MAD) that protruded from the tail of the aircraft.

"P-3 school was tough, the toughest four months I've ever had," he admits. "We had to learn every system on that airplane."

SITTING ON THE ENGINE COWLING OF A P-3 ORION

Still, he graduated from the class and was assigned to the unit, VP-49, a.k.a. 'The Woodpeckers,' and it was here that the Navy began delivering on its promise to, 'See the World!'

He was able to see Europe, and traveled extensively wherever a need arose to work on the P-3.

Breland mustered out in 1986 and immediately sought work as an aircraft mechanic in the private sector. He challenged the FAA board in Atlanta, and easily garnered his commercial A&P license.

From there, it was jobs at Sikorsky Helicopters, Lockheed (where he briefly worked on the space shuttle), Boeing, and finally, Southwest Airlines where he has worked for over 32 years, eventually attaining the position of Head Mechanic. He never regrets joining the Navy and the experience and job skills it provided to him.

He lives in Gallatin, Tn. with his wife, Jaqueline, and their two German Shepards, Luke, and Max. He has two daughters, and two grandchildren.

A SOLDIER'S POEM

THOMAS F. BALDWIN, STAFF SERGEANT, U.S. ARMY

The first time Daniel, Michael, and Randal Baldwin heard their father speak of being in the war was when they were huddled around the black and white TV in their living room watching the iconic show, “Combat.”

Until that evening, Thomas Baldwin had remained stoically reticent about his time spent serving his country. This time however, he couldn’t keep his opinion to himself.

“Street fighting is one of the dirtiest, filthiest things I’ve ever seen in my life,” he told his sons bluntly.

Randal, the youngest of his sons, gestured to the violent scene playing on the TV and probed deeper. “So, how real is this?”

Tom Baldwin smirked and said, “Eh, it’s okay…it’s TV.”

TV was a long way from becoming a ubiquitous commodity when in 1923, Tom, then only one year old, and the rest of his family travelled by Conestoga wagon to escape a Cholera epidemic in Kirkland, Wa. to return home to their native

Minnesota. Tom rode in a basket seat on the back of the wagon while the other children walked some 1,500 miles. The trip took over a year.

After settling in St. Paul, Mn. Tom's teen years during the Great Depression were taken up with trying to find employment of any stripe. He set bowling pins and hauled beef hides, before eventually getting a job as an apprentice butcher at Hormel. He was working at Hormel when he married Betty Flynn in June 1940.

Thomas enlisted in the U. S. Army some time in 1942 – 43 and spent boot camp at Ft. Leonard Wood, Mo. before being assigned to the 28th Infantry Division-110th Infantry Regiment (They were known as the *Bloody-Bucket-Boys* to the Germans). He was shipped to France six weeks after the D-Day invasion of Europe in July, 1944. Sergeant Baldwin served in Normandy, Northern France, Belgium, the Rhineland, and the Ardennes in Central Europe before being wounded in the Hurtgen Forest, the longest battle on German ground during World War II and the longest single battle the U.S. Army has ever fought.

WITH SOME SQUAD MEMBERS IN EUROPE

Sergeant Baldwin had luck on his side that day. Peeking up from a foxhole just shoulder high, he raised his head just enough that a sniper who would have had his head in his sights pulled the trigger and missed the head-shot. The bullet left a scar on the underside of his chin and entered his left arm.

He was evacuated from the front and returned home. As he recovered from his wounds stateside, the German's surrendered and he was offered an 'opportunity' to serve in the Pacific in exchange for getting a bump in rank to 2nd Lieutenant. Since he had recently lost his brother fighting in the

Philippines and had enough points to retire, he left the service in 1945 as a highly decorated Staff Sergeant.

Thomas and Betty moved from Minnesota to Southern California in 1967 along with their three children, eventually settling in Gardena, Ca. For years he suffered from severe headaches due to his wounds suffered in battle. He also had bouts of – then undiagnosed or understood, PTSD, sometimes hearing a plane fly overhead in the middle of the night and dragging his wife off the bed while covering her and screaming, "Incoming!"

His headaches were eventually diagnosed and he lived pain-free until dying quietly from arteriosclerosis while at the Los Angeles Airport on February 11th, 1984.

Besides his ailments, he did return home from the war with some souvenirs: a German Luger Pistol – that unfortunately got stolen from his sea-bag, and this poem from his time spent in battle:

A Letter to Mom

Brought home by Tom Baldwin from the Battle of the Bulge

Author Unknown

I'm down in a hole made by a shell
The guns around me are making it hell.
With the flash of a gun I can see a lad,
who should be home with his Mom and Dad.

By the flash of another, I can see his face
It's white and drawn, as if made of paste.
He's coming my way now. It's he or I.
Neither one of us want to die.

I shot him Mom. He fell with a scream.
This war is worse than my craziest dream.
I'm praying to God to give me strength,
through this terrible war that has no length.

He was a good-looking boy. About twenty-one
It was I who killed him! Lord. The harm being done.
I was trained to do so, there's a war to be won.

STEPPING UP TO THE PLATE

PAT WEBB, SEAMAN, U.S. NAVY

Pat Webb sat across from Knox Doss, a thirty-year military veteran, and principal of Vena Stuart School in Gallatin, Tn. Doss was the man who would decide the next phase of Webb's life. On the desk between them was Webb's resume, neatly typed and proofread. Webb, smartly dressed, sat ramrod straight, wanting to make a good impression. Then, the most important question in the brief interview was delivered.

"Have you ever been in the military Mr. Webb?"

"Yes, sir. I was in the navy."

"Then you're hired," Doss said simply.

And with that, Pat Webb earned a position as a PE teacher at the school, and would go on to coach boys' and girls' basketball, as well as help out with football. It was one of many benefits his time in the service would provide him.

Born in Erin, Tn. to what he described as 'simple country' folks, he volunteered to join the Navy in 1951 at the age of 19. The Korean war had started just a year earlier, but conflict and the call to duty was nothing new in the Webb household; his father had fought in WWI, and his older brother in WWII.

He shipped out to San Diego for what would be a tough six weeks of bootcamp. Webb remembers painfully the first morning.

"We had arrived at 2 a.m. and they let us sleep…but only until 5 a.m. when they rousted us out of bed!"

The food was bad as well, and Webb remembers it as being some of the vilest smelling and tasting fare he had ever had. Three days later though, famished from all the physical activity, he had grown to savor it.

The other thing Webb had a taste for – actually, a PASSION, was baseball! And it was here in bootcamp that he had a chance to join the local Navy team. But, in this case; he wouldn't make the cut.

"They were almost all pros, those guys. They were good!" Webb recalls admirably.

After graduating bootcamp and thirty days leave at home, he was on his way again, to Guam. He was assigned as a Yeoman, and had any number of administrative and clerical duties. Here, in the tiny island territory in the Western Pacific, he was to experience heat and humidity unlike any he had ever seen in middle Tennessee. And the torrential rain was even worse.

Webb recalls his first time watching a movie in the base outdoor amphitheater.

"I see all these guys coming to the movie carrying rain ponchos and I'm wondering; 'what's that all for? There's hardly a cloud in the sky.'"

A few minutes later, a single small cloud, hardly threatening, appeared moving toward them and all the guys start putting their ponchos on. Then the sky opened up.

"I never went to the movies without my poncho again, "Webb laughs.

When he eventually adapted to the heat and humidity, Webb tried out for one of the baseball teams on the island. Again, they were a very talented group of AAA and pro players and he wondered if he could make the team. It was only after he observed the team's catcher that he saw a weakness he could exploit.

"He didn't have an arm," Webb recalls. "And I had a rocket arm!"

And so, the veteran 2nd baseman sold himself as a catcher, tried out, and made the team. It was a great time while it lasted.

18 months later he would be transferred to Hawaii, but not before heading north to the volatile waters off of Korea.

"We were put onto an LST, (landing ship tank), one of the worst boats you could be on, and were sent to patrol waters off of the Korean coast."

One time was especially hairy and Webb remembers it as the scariest moment of his life.

"We headed out of Pusan Harbor late at night, in rough seas. I was with guys down below watching a movie when all of a sudden, the front of the ship went straight up. We had hit a huge wave head on. And when we came down, we hit another wave and it cracked the blunt end of the ship. It was the most scared I had ever been."

Thankfully, the LST made it back to the harbor and was repaired. There were no deaths or serious injuries.

After the Korea patrol, Webb was finally shipped off to Naval Air Station, Barber's Point in Hawaii, a duty station he considered it to be a paradise compared to Guam. Here, he also played on the baseball team and was the only non-pro player on the team.

He spent the final 2 years of his military service here and was to be mustered out in 1955 to return to a normal life. But fate had different plans, and would conspire to have him experience one last harrowing adventure.

"We were loaded onto a MARS amphibious aircraft, and headed back to San Francisco. It was the largest allied flying boat to enter production. (Only Howard Hughes Spruce Goose was larger) and had been problematic."

A short time after takeoff, one of the four engines failed and they had to turn back to Hawaii. After repairs they took off again and got further this time, but then another engine failed. They were just about at the point of no return so the pilot decided to risk it and continue on. They limped into San Francisco Bay several long hours later, alive but shaken.

THE MARS AIRCRAFT

After his military service and subsequent employment at Vera Stuart, Coach Webb continued on with an illustrious career in education and coaching. He worked at Gallatin Jr. High for 5 years, and then at Gallatin High School for a whopping 23 years as a beloved teacher and coach. He was a man who made a positive difference in hundreds of students lives.

He still lives with his wife in Gallatin, Tn. and they have 3 children, 5 grandchildren, and 1 great-grandchild.

BROTHERS IN ARMS

BUDDY & WAYNE VANTREASE, SPEC – 4, U.S. ARMY

WAYNE AND BUDDY TODAY

They grew up the sons of Roy and Gladys Vantrease in Gallatin, Tn. working on the family's 1600-acre farm.

The eldest son, Buddy, had graduated high school the year before and, at the urging of his parents, enrolled in Tennessee Tech, where he immediately felt out of place. Soon after, he returned to the farm where he remembers getting his draft notice in the mail.

"I really didn't know what it was…or what it meant," he admits today.

He found out pretty quick.

The year was 1966, and the Viet Nam War was just beginning to ramp up.

A couple of weeks later, he went to the Gallatin Post Office along with about twenty other local boys to begin the induction process. He was immediately put in charge of the group by several MP's that were on site.

"If someone doesn't get on the bus when it arrives, you let us know," they barked at him.

Eventually, they all climbed onto the bus and headed south to Fort Benning, Ga.

"I didn't really know where I was," Buddy recalls. "But it was the hottest place I had ever been!"

Meanwhile, back on the home-front, Wayne missed his older brother, but was enjoying some new-found freedoms as well.

"I kind of liked it in a way," he admits sheepishly. "I had the bedroom all to myself, and I could drive our only car pretty much whenever I wanted."

He adds, "I guess I was kind of selfish."

Despite the intense heat, bootcamp for Buddy was easy and unremarkable. Before long he had graduated and was on his way to AIT (Advanced Individual Training) school to become an MP.

Graduation from AIT was quickly followed by deployment to Pleiku, in the central highlands of Viet Nam, where his duties consisted of maintaining road security.

"I was basically like a Highway Patrol Officer," he explains. "Except the stakes were much higher."

IEDs and mines planted by the enemy were commonplace, and on more than one occasion Buddy witnessed the shocking aftereffects of some of them.

SPEC – 4 BUDDY VANTREASE

Unglamorous as his job was, it sometimes had some unique 'perks' that came with the authority of managing a road that vehicles had to travel on.

"We needed some lumber to build a shelter for our motor-pool," Buddy explains. "But only the engineers could get building materials."

Then, good fortune smiled on them.

"A military tractor-trailer rig was coming down the road — and it was loaded with lumber!"

Buddy and the other MPs stopped the rig and told the driver that it was an *'unsecured load and that they could not let it travel any further down the highway.'*

"We offloaded it right then and there, and the tractor and empty trailer continued on down the road."

Buddy's unit magically got the wood for their new motor-pool shelter.

Highway robbery notwithstanding, there was a war going on, and with real dangers.

During the Tet Offensive of 1968, their compound went on high alert and essentially hunkered down for the coming onslaught. The only problem for Buddy, was that he had no weapon with which to defend the compound – or himself! No sidearm, no M-16… Nothing!

His lack of proper defensive weapons went on for about two agonizing weeks, before he became so fed up, he wrote a letter explaining the situation to his uncle, a high-ranking member of the USAF at the Pentagon. Soon afterwards, Buddy Vantrease had a weapon.

After surviving the Tet Offensive FUBAR, Buddy returned stateside to Fort Campbell, Ky. where he continued his MP duties before mustering out in July 1968.

He returned to Gallatin and to the farm, even purchasing a Deuce and Half for himself.

"I always had an affinity for the beastly vehicles," he explained. "And I got to drive it in the Veteran's Day Parade in Gallatin for several years."

Eventually, he took a job with the post office where he worked for thirty-four years before becoming a courier for the Commerce Union Bank.

Retired, he still lives on the farm he grew up on. He has two sons and three grandchildren.

When Buddy returned home, his younger brother, Wayne was still in high school. Not the same high school he was in when his older brother went off to fight the war, but one of many he had cycled through due to disciplinary issues.

"I had a lot of trouble with authority," Wayne says unabashedly. His attitude issues were about to change however — courtesy of Uncle Sam; he received his draft notice not long after he graduated.

A couple weeks later he was on a bus to Fort Campbell, Ky. to attend basic training. It was a moment that would change his life.

"I saw right there and then what real authority was like…and I knew that I couldn't buck the system like I had done in school."

And so, he became a model recruit. He scored well on academic testing and, like his older brother, Buddy, had no issues with the physical demands.

Despite the fact he had no prior medical experience, he was designated to become an Army medic and was sent to AIT in San Antonio as soon as he graduated.

"Although I felt good about the Army seeing enough in me to make me a medic, I was disappointed in the training, which seemed too much like a hospital setting as opposed to the field."

He shipped out to An Khe, Viet Nam and thought the stars had aligned well for him. He worked at an aid station, mostly handling 'sick call' issues and other routine duties. His good fortune was not to last.

"Get up! Get up!" the sergeant shook him awake at 4:30 in the morning. "You have to be on the chopper pad in twenty minutes!"

He was going to the field, and like his brother during Tet, had no weapon.

"Here," the Sergeant said, shoving a .45 into his hands. "Now, get going!"

Dressed, and with weapon in hand, he left his cushy position and headed into unknown territory. Worst of all, it was on a helicopter; something he had never experienced in his life.

When he arrived at the chopper, there was no room inside. He asked somebody where he was supposed to sit and was told to *"park his butt on the edge, put his boots on the skids, and hang on for dear life!"*

Hang on he did, but it was almost not enough as the chopper climbed, then made a hard bank towards his side. He was nearly about to fall out when someone behind him grabbed his rucksack at the last second and held him in. He thought for sure his parents back home would be getting the dreaded DOD letter.

He finally landed - shaken but alive, at *LZ (landing-zone) Challenge,* where he soon found himself fitting in and being treated very well by the grunts.

"They came to me for everything," Wayne explains. "Not just physical stuff, but emotional things like wanting to talk about a 'Dear John' letter they had received."

The military has an expression that, 'the best ship or base you were ever on…was the last one you were on.' *LZ Challenge* was followed a short time later by a transfer to *LZ Hard Time*. It was here that the younger Vantrease had his first experience with incoming rifle fire – although he could never see the perpetrator.

"We had an enemy sniper who liked to mess with us," he explains. "He must have been several thousand yards away because we never saw him, plus he never hit anybody. You could just be standing there, smoking a cigarette or whatever, and a bullet would hit the ground at your feet. It really kept you unnerved."

It was at *LZ Hard Time* that he also experienced the lowest point during his tour in country, as well as the best.

"I was coming up on my twenty-first birthday," he explained. "And I was really depressed that I would be spending it here in a war-zone as opposed to partying back home. Plus, it was the monsoon season and we weren't getting any mail from home — which really kept me going."

SPEC – 4 WAYNE VANTREASE AT LZ HARD TIME

Then, on his birthday, the mail truck pulled in and it had a big package for him from home. He opened it and couldn't believe what he saw.

"Marshmallows," he says incredulously. "It was filled with marshmallows!"

He didn't really care for the sugary snacks, but others in the unit did, and so they started scarfing them up.

It was then, like an archeologist discovering a buried artifact, did he realize that the pillowy treats were only the cushioning for the real surprise inside —a homemade cake for his twenty-first birthday!

"It was the best moment I ever had in Viet Nam!" he recalls.

LZ Hardtime was followed by a transfer to a big aid station in Chu Lia. It wasn't the best of transfers or, for the best of reasons; Vantrease was needed because most of the other medics had been killed a couple of days prior in a big rocket attack.

'The best ship or base you were ever on...'

He was at Chu Lia for about 3-4 months before being put on a Deuce and a half for a grueling 20-hour ride further north to the DMZ – which he describes as the scariest time he had ever spent in-country.

It was also the most uncomfortable as the temps got very cold at night and in a strange twist, brought mosquitos out in even bigger droves, which at times completely covered any exposed parts of your body.

'The best ship or base you were ever on...'

He mustered out of the Army as a Spec-4 in October of 1971 and landed on US soil in Tacoma, Wa. Here, he was ushered with hundreds of other returning service members into a big auditorium where they were given the sad truth about the state of the country.

"You will find a lot of changes have happened in your absence," the speaker began. *"And do not think you are returning home as a hero. Because, in most people's eyes, you are definitely <u>NOT</u> A HERO."*

Wayne Vantrease returned to the family farm, kept his mouth shut about his service and maintained a low profile. Eventually, he went into sales, and then finally, the mortuary business. He retired twelve years ago and has three children and six grandchildren. He also lives on the family farm in Gallatin Tn. not far from his brother, Buddy.

The warnings he received from the speaker in Tacoma about not feeling like a hero continued to haunt him for a long time until a fateful day several years ago at a local elementary school. At his granddaughter's insistence, he participated in a veteran appreciation day at the school.

"We walked down the hallways of the school and on either side were all of the students and staff," he recalls wistfully. "They each had an American flag in their hands and they waved it and cheered us on as we walked by."

He adds wistfully, "It was the first time in my life, I ever felt like a hero."

BY THE STROKE OF A PEN

CHARLIE SHARP, PETTY OFFICER 3RD. CLASS, U.S. NAVY

It was nearly thirty years after William Butler Yeats wrote, “The Coming of Wisdom with Time” that the impact of those words would echo in Charlie Sharp’s head.

For the moment though, on December 7th, 1941, all the Inglewood, Tn. native could think about was the flames and smoke billowing into the sky as the Jere Baxter Middle School burned in front of him. A recent graduate of the school, Sharp joined the on-lookers that had gathered to watch the inferno. Then, word came that Pearl Harbor had been bombed by the Japanese. The world was suddenly a lot different.

“I knew plenty of older guys that I had gone to school with who were in the Navy,” Sharp recalls. “And I only hoped that they would survive.”

Many, did not.

Only a high school freshman at the time, he continued with education until one week before his eighteenth birthday when he began lobbying his father to sign the papers that would allow him to join the Navy early.

His father eventually agreed, and before Sharp knew it, he was on his way to Great Lakes, Il. for several months of boot camp. Sharp remembers the experience as being tough, with an emphasis on discipline. One time in particular sticks out as he got in big trouble for not wearing his cap.

“You would have thought they was gonna hang me,” Sharp laughs about it now.

Since his father was a pharmacist and Sharp had worked for him in his drug store in town, he had a better than average knowledge of medicine. This talent was not lost on the War Department and so, after graduating boot camp, he was shipped off to San Diego to be trained as a medical corpsman.

Several months after graduating, he was sent to work at the Navy Hospital in Dublin, Ga. Being only one state over allowed Sharp to sometimes take quick trips home, but that luxury ended after only several months when his new orders came in to head to Okinawa.

"We shipped out of San Francisco three days after the first atomic bomb had been dropped on Hiroshima," Sharp remembers ironically.

After a thirty-day trip of island hopping, they eventually reached their duty station. But by then, Sharp and the others of his unit knew that Japan had surrendered, and the war was finally over.

What he didn't realize until the war was over was that half of his unit had been earmarked to be assigned to the First Marine Division that had been staging in China for an all-out assault on the island of Japan. Sharp would have been part of the first wave in the initial invasion of the island, an assault that many historians said could have been some of the bloodiest fighting and cost the lives of millions.

It was after learning the potential fate of his son that Sharp's father said to him one day, "Boy, I'm gonna tell you something about Harry Truman; people may complain about him, but he saved your life with the stroke of a pen. Do you realize that?

It took a while for the ramifications of that historic decision to sink in, but eventually, Charlie Sharp realized that his father was correct. It was the *'coming of wisdom with time.'*

He finished his stint in the Navy by working at various hospitals on Guam and after mustering out, returned home and worked at his father's drug store while attending college where he became a pharmacist.

Charlie Sharp was father to 2 daughters (losing one), and still has 4r grandchildren, and a whopping 13 great-grandchildren. He is proud of the contributions he made in WWII, and considers his time spent to have been a worthwhile experience. He lives in Gallatin, Tn.

ABOUT THE AUTHOR

Christopher J. Lynch is the author of many works including: the **ONE EYED JACK** crime novel series about a professional blackmailer with a heart, **DARK STATE**, a modern-day thriller about a terrorist attack that takes down the California electrical grid, and

EDDIE: THE LIFE AND TIMES OF AMERICA'S PREEMINENT BAD BOY, the biography of Ken Osmond, who played the conniving Eddie Haskel on the iconic TV series, Leave it to Beaver.

He and his wife, Charlotte, are also Gold Star parents, losing their beloved son as he proudly served in the United States Marine Corp. They live in Tennessee with their spirited puppy, Sugar.

To read more about Christopher and his other books, visit his website: www.christopherjlynch.com

Made in United States
North Haven, CT
10 November 2026

22095603R00124